METHODS IN MOLECULAR BIOLOGY™

Series Editor
John M. Walker
School of Life Sciences
University of Hertfordshire
Hatfield, Hertfordshire, AL10 9AB, UK

For other titles published in this series, go to
www.springer.com/series/7651

Neural Cell Transplantation

Methods and Protocols

Edited by

Neil J. Scolding and David Gordon

University of Bristol Institute of Clinical Neurosciences, Frenchay Hospital, Bristol, UK

 Humana Press

Editors
Neil J. Scolding
University of Bristol Institute of Clinical
  Neurosciences
Frenchay Hospital
Bristol, UK

David Gordon
University of Bristol Institute of Clinical
  Neurosciences
Frenchay Hospital
Bristol, UK

ISBN: 978-1-60327-930-7   e-ISBN: 978-1-60327-931-4
ISSN: 1064-3745   e-ISSN: 1940-6029
DOI: 10.1007/978-1-60327-931-4
Springer Dordrecht Heidelberg London New York

Library of Congress Control Number: 2009920071

Cover illustration: Derived from figure 1E in chapter 7

Printed on acid-free paper

Springer is part of Springer Science+Business Media (www.springer.com)

Preface

Few areas of neurobiology or of clinical neuroscience have changed so dramatically over the past decade as that of neural transplantation, and overwhelmingly the most powerful contributor to this change has been the stem cell: neural, embryonic, and other. Our whole approach to cell transplantation as a therapeutic approach to neurodegenerative and other neurological disease, and indeed about spontaneous repair in the nervous system, has been radically influenced by our rapidly growing knowledge both of endogenous stem cells and their response to insult, and of exogenous stem and precursor cells and their reparative capacity. Not only have these areas changed very substantially, this area of medical and biological endeavour has also quantitatively expanded enormously over this time with the emergence of numerous new research groups contributing to this fast changing field. All these groups and their individual researchers are of course dependent for their success on sound techniques, allowing them accurately to reproduce or advance others' studies and permitting their own findings to inform and enhance other groups' research.

None of us are immune from the competitive instinct, which of course drives science further and faster forward, but what has been profoundly encouraging in the preparation of this practical book of recipes has been the extraordinarily generous willingness amongst the leaders of this field not only to share their technical skills, but to give of their time to prepare these manuscripts. We are immensely grateful to all the authors whose texts follow for allowing this book to be put together, and we hope sincerely that their efforts will allow new generations of neuroscientists to enter and contribute to this uniquely inspiring field.

Bristol, UK   *Neil J. Scolding*
David Gordon

Contents

Preface. . . . . *v*

Contributors. . . . . *ix*

PART I: INTRODUCTORY REVIEWS

1. Neural Transplantation and Stem Cells 3
    Mahendra S. Rao and Mohan C. Vemuri

2. Adult Stem Cells for the Treatment of Neurological Disease 17
    C.M. Rice and N.J. Scolding

3. Human Trials for Neurodegenerative Disease 33
    Claire M. Kelly, O.J. Hundley, and A.E. Rosser

PART II: PRACTIAL/METHODOLOGICAL CHAPTERS

4. Differentiation of Neuroepithelia from Human
    Embryonic Stem Cells. 51
    Xiaofeng Xia and Su-Chun Zhang

5. Derivation of High-Purity Oligodendroglial Progenitors 59
    Maya N. Hatch, Gabriel Nistor, and Hans S. Keirstead

6. Flow Cytometric Characterization of Neural Precursor Cells
    and Their Progeny. 77
    Preethi Eldi and Rodney L. Rietze

7. Isolation, Expansion, and Differentiation of Adult Mammalian
    Neural Stem and Progenitor Cells Using the Neurosphere Assay 91
    Loic P. Deleyrolle and Brent A. Reynolds

8. Human Mesenchymal Stem Cell Culture for Neural Transplantation 103
    David Gordon and Neil J. Scolding

9. Umbilical Cord Blood Cells. 119
    Jennifer D. Newcomb, Alison E. Willing, and Paul R. Sanberg

10. Animal Models of Neurodegenerative Diseases 137
    *Wendy Phillips, Andrew Michell, Harald Pruess,
    and Roger A. Barker*

11. Animal Models of Multiple Sclerosis 157
    Roberto Furlan, Carmela Cuomo, and Gianvito Martino

12. Transplantation of Oligodendrocyte Progenitor Cells in Animal
    Models of Leukodystrophies 175
    Yoichi Kondo and Ian D. Duncan

13. A Rat Middle Cerebral Artery Occlusion Model and Intravenous
    Cellular Delivery 187
    Masanori Sasaki, Osamu Honmou, and Jeffery D. Kocsis

14.   MR Tracking of Stem Cells in Living Recipients 197
      Eva Syková, Pavla Jendelová, and Vít Herynek

15.   Identifying Neural Progenitor Cells in the Adult Brain 217
      Stephen Kelly, Maeve Caldwell, Matthew P. Keasey,
      Jessica A. Cooke, and James B. Uney

16.   Immune Ablation Followed by Autologous Hematopoietic
      Stem Cell Transplantation for the Treatment of Poor Prognosis
      Multiple Sclerosis 231
      Harold Atkins and Mark Freedman

Index *247*

Contributors

HAROLD ATKINS • *The Ottawa Hospital Blood and Marrow Transplant Program, Ottawa, Ontario, ON, Canada*

ROGER A. BARKER • *Cambridge Centre for Brain Repair, University of Cambridge, Cambridge, UK, and Department of Neurology, Addenbrooke's Hospital, Cambridge, UK*

MAEVE CALDWELL • *Henry Wellcome Laboratories for Integrative Neuroscience and Endocrinology (LINE), Clinical Sciences South Bristol, University of Bristol, Bristol, UK*

JESSICA A. COOKE • *Henry Wellcome Laboratories for Integrative Neuroscience and Endocrinology (LINE), Clinical Sciences South Bristol, University of Bristol, Bristol, UK*

CARMELA CUOMO • *Neuroimmunology Unit – DIBIT and Department of Neurology, San Raffaele Scientific Institute, Milan, Italy*

LOIC P. DELEYROLLE • *Queensland Brain Institute, The University of Queensland, Brisbane, QLD, Australia*

IAN D. DUNCAN • *Department of Medical Sciences, School of Veterinary Medicine, University of Wisconsin-Madison, Madison, WI, USA*

PREETHI ELDI • *Queensland Brain Institute, The University of Queensland, Brisbane, QLD, Australia*

MARK FREEDMAN • *The Ottawa Hospital Multiple Sclerosis Research Clinic, The Ottawa Hospital, Ottawa, ON, Canada*

ROBERTO FURLAN • *Neuroimmunology Unit –– DIBIT and Department of Neurology, San Raffaele Scientific Institute, Milan, Italy*

DAVID GORDON • *University of Bristol Institute of Clinical Neurosciences, Frenchay Hospital, Bristol, UK*

O. J. HANDLEY • *Brain Repair Group, School of Biosciences, Cardiff, UK*

MAYA N. HATCH • *Department of Anatomy and Neurobiology, Reeve-Irvine Research Center, University of California at Irvine, Irvine, CA, USA*

VÍT HERYNEK • *MR-Unit, Department of Radiodiagnostic and Interventional Radiology, Institute for Clinical and Experimental Medicine, Prague, Czech Republic*

OSAMU HONMOU • *Department of Neurology and Center for Neuroscience and Regeneration Research, Yale University School of Medicine, New Haven, CT, USA*

PAVLA JENDELOVÁ • *Institute of Experimental Medicine ASCR, Prague, Czech Republic*

MATTHEW P. KEASEY • *Henry Wellcome Laboratories for Integrative Neuroscience and Endocrinology (LINE), Clinical Sciences South Bristol, University of Bristol, Bristol, UK*

CLAIRE M. KELLY • *Brain Repair Group, School of Biosciences, Cardiff, UK*

STEPHEN KELLY • *Henry Wellcome Laboratories for Integrative Neuroscience and Endocrinology (LINE), Clinical Sciences South Bristol, University of Bristol, Bristol, UK*

HANS S. KEIRSTEAD • *Department of Anatomy and Neurobiology, Reeve-Irvine Research Center, University of California at Irvine, Irvine, CA, USA*

JEFFERY D. KOCSIS • *Department of Neurology and Center for Neuroscience and Regeneration Research, Yale University School of Medicine, New Haven, CT, USA*

YOICHI KONDO • *Department of Medical Sciences, School of Veterinary Medicine, University of Wisconsin-Madison, Madison, WI, USA*

GIANVITO MARTINO • *Neuroimmunology Unit – DIBIT and Department of Neurology, San Raffaele Scientific Institute, Milan, Italy*

ANDREW MITCHELL • *Cambridge Centre for Brain Repair, ED Adrian Building, Cambridge, UK*

JENNIFER D. NEWCOMB • *Department of Neurosurgery, Center of Excellence for Aging and Brain Repair, University of South Florida College of Medicine, Tampa, FL, USA*

GABRIEL NISTOR • *Department of Anatomy and Neurobiology, Reeve-Irvine Research Center, University of California at Irvine, Irvine, CA, USA*

WENDY PHILLIPS • *Cambridge Centre for Brain Repair, ED. Adrian Building, Cambridge, UK; Department of Neurology, Addenbrooke's Hospital, Cambridge, UK*

HARALD PRUESS • *Priller Laboratory, Department of Neurology, Charite University Medicine, Berlin, Germany*

MAHENDRA S. RAO • *Invitrogen Corporation, Grand Island, NY, USA*

BRENT A. REYNOLDS • *Queensland Brain Institute, The University of Queensland, Brisbane, QLD, Australia*

C. M. RICE • *University of Bristol Institute of Clinical Neurosciences, Frenchay Hospital, Bristol, UK*

RODNEY L. RIETZE • *Queensland Brain Institute, The University of Queensland, Brisbane, QLD, Australia*

A. E. ROSSER • *Department of Neurology and Medical Genetics, School of Medicine, University of Cardiff, Cardiff, UK*

PAUL R. SANBERG • *Department of Neurosurgery, Center of Excellence for Aging and Brain Repair, University of South Florida College of Medicine, Tampa, FL, USA*

MASANORI SASAKI • *Department of Neurology and Center for Neuroscience and Regeneration Research, Yale University School of Medicine, New Haven, CT, USA*

NEIL J. SCOLDING • *University of Bristol Institute of Clinical Neurosciences, Frenchay Hospital, Bristol, UK*

EVA SYKOVÁ • *Institute of Experimental Medicine ASCR, Prague, Czech Republic*

JAMES B. UNEY • *Henry Wellcome Laboratories for Integrative Neuroscience and Endocrinology (LINE), Clinical Sciences South Bristol, University of Bristol, Bristol, UK*

MOHAN C. VEMURI • *Invitrogen Corporation, Grand Island, NY, USA*

ALISON E. WILLING • *Department of Neurosurgery, Center of Excellence for Aging and Brain Repair, University of South Florida College of Medicine, Tampa, FL, USA*

XIAOFENG XIA • *Departments of Anatomy and Neurology, School of Medicine and Public Health, Waisman Center, University of Wisconsin-Madison, Madison, WI, USA*

SU-CHUN ZHANG • *Departments of Anatomy and Neurology, School of Medicine and Public Health, Waisman Center, University of Wisconsin-Madison, Madison, WI, USA*

Part I

Introductory Reviews

Part

Chapter 1

Neural Transplantation and Stem Cells

Mahendra S. Rao and Mohan C. Vemuri

Summary

Recent results have raised important questions on our ability to amplify stem cell populations in sufficient numbers as to be useful for therapy. Several reports have indicated that human stem cell populations harvested from the adult have low or undetectable telomerase levels, age in culture, and may not be propagated indefinitely. Other groups have shown that stem cells age and as such, their properties will have changed depending on the age of the individual from which they are harvested, and the time for which they are propagated in culture. Other groups have shown that cells maintained in culture may undergo alterations as they are propagated, and that these alterations may alter the predicted behavior of stem cells. Yet others have shown that human cells differ from their counterparts in other species in significant ways and have identified important difficulties in assessing cells in a xeno environment. Clinical colleagues have identified issues of variability and difficulties in the long-term follow-up that is being requested. Researchers in the stem cell field focused on translational work need to develop a practical plan that takes into account such difficulties while developing manufacturing protocols, designing animal studies, or developing trial protocols. Such proactive planning will be critical in ensuring a successful transition from the bench to the clinic.

Key words: Neural stem cells, Stem cell, Transplantation, Cell therapy, Clinical trials, Immune response, Adult stem cells, Animal models

1. Introduction

Cells irrespective of the investigators preference of type (stem, progenitor, or differentiated cell) are thought to help enhance repair in the nervous system by one or a subset of the mechanisms described (**Table 1**). Based on their proposed mechanism several practical issues become important in evaluating how cell therapy will be performed *(12)*. These practical issues may include important

Neil J. Scolding and David Gordon (eds.), *Methods in Molecular Biology, Neural Cell Transplantation vol. 549*
DOI: 10.1007/978-1-60327-931-4_1, © Humana Press, a part of Springer Science+Business Media, LLC 2009

Table 1
Methods of neural differentiation

Methods	Refs
Direct differentiation	*(1–3)*
EB formation	*(4)*
RA treatment	*(2, 5)*
Growth factor withdrawal	*(5, 6)*
Coculture with astrocytes of PA-6	*(7, 8)*
Stochastic differentiation followed by selection using cell surface epitopes	*(9–11)*

decisions as to whether cell type A or B is preferred, what dose and at what stage in the disease process intervention should be contemplated, what safety studies should be performed, and what readouts of efficacy are most suitable. Subsequent chapters in this book provide detailed methodologies covering specific cell types, methods of analysis, protocols for assessment of cells, and methods of inserting cells to their specific targets.

In this introductory chapter we will not discuss a particular disease model in detail or provide detailed protocols but rather provide certain general guidelines that appear to have a reasonable consensus. In general there is no "one size fits all" cell for therapy and there is no optimal generalized method of cell delivery. However, there are certain generalized issues that one has to face irrespective of the cell type chosen and the therapeutic target. These are discussed briefly in subsequent sections.

2. Autologous Versus Allogeneic Cell Use for Therapy

Beginning from the 1990s, researchers have actively begun to search for possible cell resources and the cell repertoire has been enlarged in the past 10 years *(13)*. Fetal cells, adult stem cells, a variety of lineage-specific precursor cells, cells transdifferentiated from other tissues or derived from embryonic stem cells (ESCs) have all begun to be used in animal studies and their characteristics have been clarified and evaluated for potential cell therapy sources. The number of cell populations used in animal model is quite remarkable and there is evidence of possible successful functional recovery in a number of disease models. There is in general no dispute among scientists of the utility of adult stem cells and their potential therapeutic role *(14)*. Indeed the only stem cells or stem cell-containing products in clinical use are those derived from adult populations. Bone marrow transplants, limbal stem cell therapy, cord blood transplants, pancreatic islet transplants, and carticell (a cartilage progenitor product) are all examples of successful adult stem or progenitor cell therapy. In contrast, while ESC from mice have been used successfully to derive transgenics and chimeras and we have known about mouse ESC since the 1980s no ESC or ESC product is currently used in the clinic *(15, 16)*.

Perhaps the most important choice that investigators need to make is to determine if one will work with autologous or allogeneic cells **(Table 2)**. Indeed, the entire therapeutic paradigm changes when one considers including immune suppression (or demonstrating that these cells are not rejected) and as such the cost, follow, and even possible therapeutic targets need to be reevaluated. In addition, the regulatory environment treats such therapies differently as well. In the USA autologous cells are governed under different sections of the relevant PHS act (with exceptions depending on whether they are manipulated in culture and if the use is homologous or nonhomologous). The EU regulations similarly regulate allogenic and autologous cell types differently. In general the regulations are much more stringent for allogeneic therapy and to our knowledge no research investigator has initiated a clinical trial with allogeneic cells *(17, 18)*. Thus, on a practical level this is the most important distinction that a researcher needs to make as he/she proceeds to the clinic.

Table 2
Problems facing investigators

Problems facing investigators	Comments
CMC issues	Xeno issues, lot-to-lot variability, release criteria, measures of potency and efficacy
Animal model issues	Is the model appropriate? Are primate models critical? Can human-to-human transplants be modeled in human to mouse experiments?
Tracking and tracing and follow-up issues	Biodistribution, tumor formation studies, labeling
Patient selection criteria	Allelic variability, long-term immune suppression, methods of delivery
Tracking and tracing and follow-up issues in patients	Noninvasive monitoring, engineering suicide genes, etc.

3. Scale-Up and Manufacturability

Regenerative medicine and cell therapy are based on the premise that large numbers of "normal" cells will be available from adult, fetal, or embryonic sources and that these cells will migrate, integrate, and survive for a sufficient length of time as to be clinically useful. It has become clear that such large numbers of cells are not available by direct harvest from a donor source but require at least limited propagation in culture. Thus, an important issue facing investigators is growing cells in sufficient numbers for treating patients in an aseptic environment with adequate safeguards, sterility, and traceability (**Fig.1**). Guidelines for the manufacture of biologic have been developed and codified under CMC (chemistry and manufacturing controls) and GMP (good manufacturing practices) guidelines. While expertise on CMC and GMP practices is widespread given the use of antibodies, viral gene delivery, and peptide and growth factor treatment there are nevertheless several additional issues with the use of cells in general and stem cells in particular *(16, 19)*.

In the case of stem cells an additional issue of stability of the population arises as it is known that the undifferentiated stem cell state is unstable and prone to change as cells are propagated in culture. Indeed, it has been extremely difficult to propagate any human stem cell population over prolonged periods. Hematopoietic stem cells (HSC) and cord blood stem cells are examples of a failure

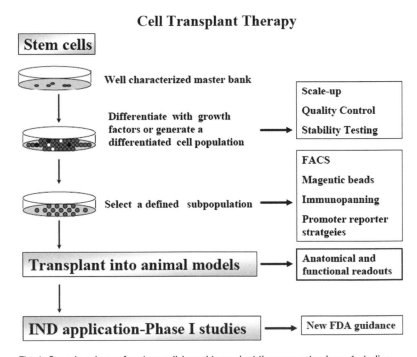

Fig. 1. Generic scheme for stem cell-based transplant therapy up to phase 1 studies.

to propagate to any degree while mesenchymal stem cells (MSC) are examples of adult stem cell with a limited proliferation potential. However, as discussed earlier, a critical requirement for cell therapy is the ability to maintain cells as a stable self-renewing population that retains the ability to differentiate into the cell type required.

It is equally important to develop a set of quality control procedures that will allow assessment of the state of the cells rapidly without using too many cells. This includes testing the major phenotypic changes that might be predictive of an alteration in cell behavior. A list of measurements we have proposed is listed in **Table 3**. Our rationale for their selection was based on the assumption that cells maintained in culture are under constant selection pressure to divide and self-renew, and that changes can occur in a stochastic manner under all culture conditions *(15, 20, 21)*. Severely detrimental mutations that do not confer a growth advantage will be lost, while those that inhibit death, accelerate growth, or alter differentiation will be selected for. In addition,

Table 3
Best characterized differences between human and rodent stem cells

Embryonic stem cells	Rodents	Humans
Growth Characteristics		
Colony morphology	Piled	Flat
Distinct Borders for colony	–	+
Population doubling time	12 h	36 h
Cloning efficiency	Good	Difficult
Cell surface and other antigens		
LIF dependence	+	–
LIF receptor-gp 130	+	–
LIF signaling inhibition by SOCSI	–	+
FGF2 dependence	–	+
FGF4	+	–
SSEA1	+	–
SSEA3	–	+
SSEA4	–	+
TRA-1–61	–	+
TRA-1–81	–	+
GCTM-2	–	+

(continued)

Table 3
(continued)

Embryonic stem cells	Rodents	Humans
CD9	+	+
CD133	+	+
OCT 4	+	+
Nanog	+	+
Sox2	+	+
β-III tubulin	–	Low levels
Integrins	$\alpha5, \alpha6, \beta1$	$\alpha6, \alpha1, \beta1, \beta2, \beta3, \beta4$
Enzyme activities		
Alkaline phosphatase	+	+
Telomerase	+	+
Neural stem cells	Rodents	Humans
Growth characteristics		
Timing of appearance	E10.5	5 weeks
Astrocyte appearance	Late	Early
Cell cycle, number of divisions	Short	Long
Cell surface and other antigens		
NGF,PDGF,CNTF, LIF	Differ	Differ
MHC class 1 expression	–	+; high level early on
CD95	–	+
Glial –A2B5 marker	–	+in cortical NPCs
Olig-2	–	–/?; likely in cultures
NG-2	–	–
Glial GFAP in NSC	– fetal NSC	–
Nestin	+	+
SOX2	+	+
O4/GALC	–	–
Oct4	–	–
Enzyme activities		
Neuron-specific enolase	–	?
Carbonic anhydrase	–	?
Telomerase	High	low

if a cell undergoes sufficient passage these changes will become fixed in the genome or epigenome and the cells will be irretrievably changed over time. Such changes and such selection have been observed with cell lines and with mouse ESC cultures, and we reason that human stem cells would also undergo such selection pressure. We therefore feel that measuring the functional ability of cells, their expression profiles of key genes (telomerase, cell cycle, key markers of the ESC state, etc.), genomic stability, and epigenome (methylation, miRNA, histone acetylation, X chromosome inactivation) would be a reasonable set of tests that need to be performed. Histone modification and DNA methylation are the most commonly measured epigenetic changes *(22–24)*; both types of analysis are labor-intensive although array-based methods in development have the potential to allow inexpensive assessment of the methylation status of hundreds of regulatory elements.

A third issue facing investigators is eliminating unwanted cells from the sample cell population. In the case of undifferentiated stem cells, it has also become evident that undifferentiated cells are often not the best choice for therapy and that appropriate phenotypes must be differentiated from them, for example, differentiating ESCs to neural or glial cell types (**Fig.2**). The population of desired cells then needs to be harvested away from contaminating undifferentiated cells and other inappropriate or undesirable differentiated cell populations.

Differentiation of hESCs into Neural cells

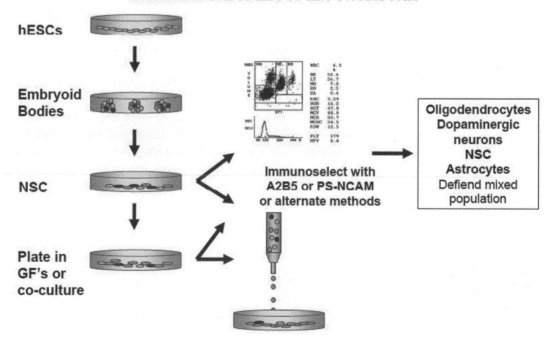

Fig. 2. Scheme for differentiation of human embryonic stem cells into neural cell types and isolation of cells needed for transplantation through cell sorting.

Cell Manufacturing Protocols

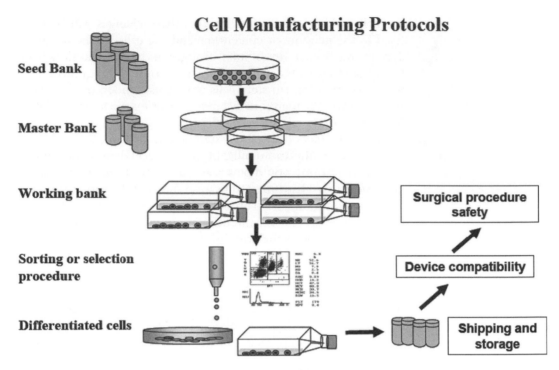

Fig. 3. Cell manufacturing protocols in cGMP and scale-up conditions.

A final issue that faces investigators is a more subtle one and that is, given the inevitable lot-to-lot variability (even if one is using stem cells) will results from a clinical trial by one group allow another to extrapolate to cells manufactured at a different site or by another company? In Battens disease will a second company's trial for neural stem cells be like what one would do for a generic drug or modification of an existing drug given the pioneering work of Stem Cell, Inc. or will any new stem cell product be treated like a new drug.

Conceptually then a manufacturing protocol consists of a master bank of undifferentiated cells that have been carefully tested to assure their quality from which working lots of cells are withdrawn, amplified to obtain sufficient numbers of cells, differentiated, and an appropriate phenotype selected based on established criteria (**Fig. 3**).

4. Determining Cell Numbers for Therapy

In addition to determining the specific cell type to be used, one will also need to determine at what stage of differentiation the cells should be transplanted and more importantly what is the dose range of cells that one will use. The decisions on the type of cell to be used, the cell numbers to be used, and the method of delivery that will be optimum will vary. For example, the number of cells to be used in Parkinson's therapy would range between 100,000 and 500,000 and would likely be delivered via a needle into the parenchyma guided by MRI *(25)*. Primarily dopaminergic neurons to the exclusion of all other cells would be important if the intent was to replace the damaged or lost dopaminergic neurons. The numbers and type of cell would be altered even for Parkinson's disease if one contemplated using stem cells or their derivatives as pumps to deliver a growth factor to enhance survival of existing neurons. In this case migration would not be a useful property and homing to a stem cell niche, for example, would be a disadvantage. In spinal cord injury, on the other hand, delivery may be parenchymal or via lumbar puncture (outpatient procedure) with a consequent difference in numbers of cell required. Since neuronal replacement would not be as important as glial replacement, the cell selected should preferably have limited neuronal differentiation (to limit aberrant neural circuit formation) and the cells should have a limited migration potential. This migratory ability should it exist, should preferably be directed toward a lesion. In stroke, given the domain is large perhaps intravenous or intra-arterial delivery may be contemplated despite the known embolic risks. In this case the ability of a cell to migrate through the blood vessel into the parenchyma and the ability to migrate to a target site may be invaluable properties.

5. Assessing Mode of Action

There are several aspects to this issue. Perhaps the most obvious is simply how does one trace cells, assess biodistribution, and quantify integrated cells. These issues while more difficult to address than assessing small molecules are nevertheless conceptually similar and we have tools and techniques to assess such issues: radioactive labeling of cells to look at short-term distribution, contrast labels to follow cells over longer time periods, and multiple methods to assess integration. However, there are also more subtle issues that one needs to consider in assessing mode of action in animal models (**Fig. 4**).

Perhaps the most important is the issue of immune suppression and evaluating this in animal models. Testing human cells in animals requires manipulating the endogenous rejection response to a xenotransplant, and thus studying immune rejection issues that may relate to transplanting allelically mismatched cells from one human donor to another cannot be readily tested.

Animal Study Issues

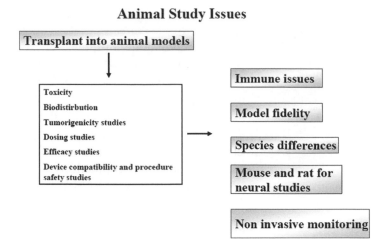

Fig. 4. Transplantation into animal models and relevant issues for clinical application.

Yet this is a critical issue, despite many reports that stem cells may not be immunogenic or may suppress a MLR reaction *(26, 27)*. It is clear that long-term survival requires some sort of immune suppression. Perhaps the only solution has been to test equivalent rat/mouse cells from a different strain in rat/mouse model. Yet even this is not very satisfying as there are many reported differences between stem cells isolated from different species *(17, 28, 29)* even if care is taken to isolate them by similar methods, at similar stages of development and maintain them in similar conditions.

The second critical issue is what we have euphemistically termed the xeno issue. Transplanting human cells in an animal model is simply not the same as transplanting human cells into human *(26)*. The reason is that no matter how similar the disease model there are a host of physiological factors that are different: the size of the cells, their cycle time, osmolarity, and endocrine differences, blood pressure and heart rate variation, and even tissue oxygen concentration. Each of these differences has cumulative effects on cells affecting their survival or efficiency of integration and has no bearing on the issue of human cells transplanted into humans. The second alternative is to consider same species cells into the appropriate model as a test for human-to-human transplants is also not optimal. Species differences between cells are well known in either marker expression, growth factor response, or sensitivity to toxins *(17)*. In addition, there is limited availability of large-scale models or equivalent cells types in many species. For example, we do not have any data that rat ESCs are available, behave the same way, or respond to the same growth factors. However, the rat is a very useful model for neural assessment. One, therefore, faces a dilemma of how one may perform realistic tests that the FDA will find appropriate.

6. Determining a Clinical End Point and Assessing Patient Recruitment Criteria

As with any clinical trial there are several issues that are common to using stem cells or a small molecule drug. These include appropriate patient selection, obtaining appropriate consent, and determining the number of patients required to obtain interpretable results from clinical trials (**Fig.5**). These are well discussed elsewhere. We focus on two issues that we feel are relatively unique to stem cell populations.

The first issue is the lot size. Generally, when autologous therapy is considered each lot is used for one patient. Even in many of the autologous therapies that are being considered lot sizes are often small and are sufficient to treat only a small number of patients *(30)*. This brings up an important issue for regulators and for investigators. How does one predict the effect and efficacy from lot to lot as there is undoubtedly variability when cells are harvested from different donors or different ages, different ethnic backgrounds, and different clinical histories and environmental exposures? An extension of this issue is determining the size of a clinical trial when such variability is likely to be confounding.

A second equally important issue that all of us face is the issue of the half-life of cells and their change after transplant. Unlike drugs that have a relatively short half-life and that are metabolized by relatively well-characterized pathways, cells are rather different. Stem cells may persist for the lifetime of an individual and for the most part respond to the environment by changing or differentiating. This differentiation is regulated by the environment, which, in most currently envisaged uses, is a nonhomologous use of a stem cell.

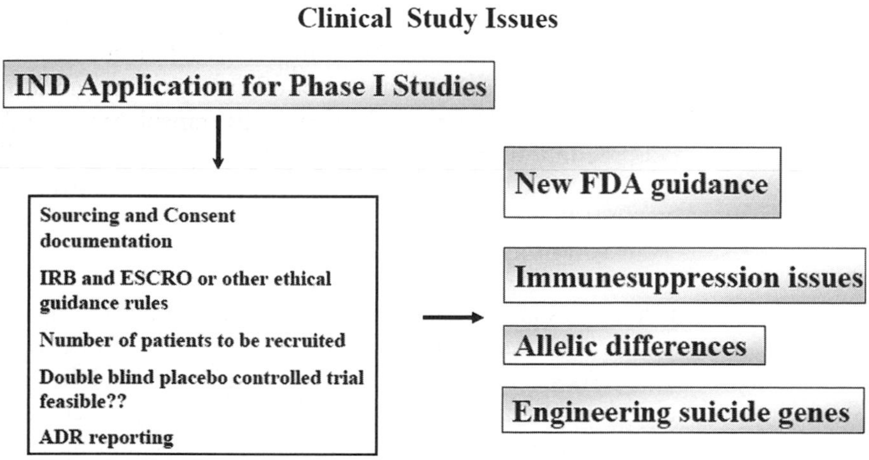

Clinical Study Issues

Fig. 5. Clinical study issues with stem cells.

A third major issue for stem cell therapy is not knowing the mechanism of action and therefore being unable to design appropriate efficacy assays and generate a dosing regime or selecting the right patients for therapy. Stem cells may integrate and replace missing cells; stem cells may mobilize endogenous repair pathways, and stem cells may provide trophic support or may do all of the aforementioned functions *(31, 32)*. This makes designing a clinical trial quite difficult as one does not quite know how many cells to transplant, how to test the quality of the cells that will predict their efficacy with reasonable efficiency, how long to follow, and how large a trial will be to have appropriate interpretable results. This has a major impact on the cost of a trial, recruitment of patients, and the ability of regulators to interpret these trials fairly. These difficulties have suggested that the standard model that we use for drug trial may not be the best for cells. Unfortunately, we do not have a better alternative that would not put patients at significant risk.

7. Translating These Concerns into Practical Solutions

Most therapeutic uses of stem cells require large number of cells maintained in a GMP facility. Likewise, for gene and drug discovery processes, it is assumed that stable and genetically identical cells will be available in large numbers. A flow chart of the choices that need to made is summarized (**Fig.1**). Perhaps the first choice is whether therapy will be autologous or allogeneic and whether amplification in culture will be required or not. If no amplification is required and the cells are minimally manipulated then processing will be straightforward and regulatory hurdles will be clear-cut. Should amplification of cells be required or cells have to be altered (transdifferentiated) for their function then even if autologous cells are used they will need to be processed in a GMP facility adhering to rigorous guidance. The emerging approach for CMC manufacture is to develop a master bank from which a working bank of cells is obtained and differentiated into an appropriate phenotype for use. Cells are selected based on cell surface epitopes using a GMPable process and a closed isolation system such as the Isolex/Dynal or Clinimacs systems. Once sufficient cells are obtained animal studies will be performed ideally with cells harvested by a GMPable process (if not in a GMP facility), and tests on the biodistribution and safety will be performed in an appropriate animal model that takes into account known species differences and the problems of transplanting human cells in an alien environment. Novel methods to follow transplanted cells and assess their biodistribution safety and efficacy will be utilized and corresponding noninvasive tests will be used in clinical trials where monitoring will be of a sufficient time period to satisfy concerns of tumorigenicity and toxicity of any kind.

References

1. Carpenter, M. K., Rosler, E. and Rao, M. S. (2003). Characterization and differentiation of human embryonic stem cells. *Cloning Stem Cells* **5**, 79–88.

2. Luo, Y., Schwartz, C., Shin, S., Zeng, X., Chen, N., Wang, Y., Yu, X. and Rao, M. S. (2006). A focused microarray to assess dopaminergic and glial cell differentiation from fetal tissue or embryonic stem cells. *Stem Cells* **24**, 865–75.

3. Zeng, X., Miura, T., Luo, Y., Bhattacharya, B., Condie, B., Chen, J., Ginis, I., Lyons, I., Mejido, J., Puri, R. K., Rao, M. S. and Freed, W. J. (2004). Properties of pluripotent human embryonic stem cells BG01 and BG02. *Stem Cells* **22**, 292–312.

4. Li, H., Liu, Y., Shin, S., Sun, Y., Loring, J. F., Mattson, M. P., Rao, M. S. and Zhan, M. (2006). Transcriptome coexpression map of human embryonic stem cells. *BMC Genomics* **7**, 103.

5. Mayer-Proschel, M., Kalyani, A. J., Mujtaba, T. and Rao, M. S. (1997). Isolation of lineage-restricted neuronal precursors from multipotent neuroepithelial stem cells. *Neuron* **19**, 773–85.

6. Miura, T., Mattson, M. P. and Rao, M. S. (2004). Cellular lifespan and senescence signaling in embryonic stem cells. *Aging Cell* **3**, 333–43.

7. Ellis, P., Fagan, B. M., Magness, S. T., Hutton, S., Taranova, O., Hayashi, S., McMahon, A., Rao, M. and Pevny, L. (2004). Sox2, a persistent marker for multipotential neural stem cells derived from embryonic stem cells, the embryo or the adult. *Dev Neurosci* **26**, 148–65.

8. Schwartz, C. M., Spivak, C. E., Baker, S. C., McDaniel, T. K., Loring, J. F., Nguyen, C., Chrest, F. J., Wersto, R., Arenas, E., Zeng, X., Freed, W. J. and Rao, M. S. (2005). Ntera2: A model system to study dopaminergic differentiation of human embryonic stem cells. *Stem Cells Dev* **14**, 517–34.

9. Shin, S., Sun, Y., Liu, Y., Khaner, H., Svant, S., Cai, J., Xu, X. Q., Davidson, B. P., Stice, S. L., Smith, A. K., Goldman, S. A., Reubinoff, B. E., Zhan, M., Rao, M. S. and Chesnut, J. D. (2007). Whole genome analysis of human neural stem cells derived from embryonic stem cells and stem and progenitor cells isolated from fetal tissue. *Stem Cells* **25**, 1298–306.

10. Rao, M. S. and Anderson, D. J. (1997). Immortalization and controlled in vitro differentiation of murine multipotent neural crest stem cells. *J Neurobiol* **32**, 722–46.

11. Zeng, X., Chen, J., Deng, X., Liu, Y., Rao, M. S., Cadet, J. L. and Freed, W. J. (2006). An in vitro model of human dopaminergic neurons derived from embryonic stem cells: Mpp+ toxicity and gdnf neuroprotection. *Neuropsychopharmacology* **31**, 2708–15.

12. Turksen, K. and Rao, M. (2005). Issues in human embryonic stem cell biology. *Stem Cell Rev* **1**, 79–81.

13. Magnus, T., Liu, Y., Parker, G. C. and Rao, M. S. (2007). Stem cell myths. Philos Trans R Soc Lond B Biol Sci., in press.

14. Ginis, I. and Rao, M. S. (2003). Toward cell replacement therapy: Promises and caveats. *Exp Neurol* **184**, 61–77.

15. Choong, C. and Rao, M. S. (2007). Human embryonic stem cells. *Neurosurg Clin N Am* **18**, 1–14, vii.

16. Zeng, X. and Rao, M. S. (2006). Human embryonic stem cells: Long term stability, absence of senescence and a potential cell source for neural replacement. *Neuroscience* **145**, 1348–58.

17. Sun, Y., Li, H., Liu, Y., Shin, S., Mattson, M. P., Rao, M. S. and Zhan, M. (2007). Cross-species transcriptional profiles establish a functional portrait of embryonic stem cells. *Genomics* **89**, 22–35.

18. Gibson, J., Ho, P. J. and Joshua, D. (2004). Evolving transplant options for multiple myeloma: Autologous and nonmyeloablative allogenic. *Transplant Proc* **36**, 2501–3.

19. Terstegge, S., Laufenberg, I., Pochert, J., Schenk, S., Itskovitz-Eldor, J., Endl, E. and Brustle, O. (2007). Automated maintenance of embryonic stem cell cultures. *Biotechnol Bioeng* **96**, 195–201.

20. Sun, Y., Li, H., Yang, H., Rao, M. S. and Zhan, M. (2006). Mechanisms controlling embryonic stem cell self-renewal and differentiation. *Crit Rev Eukaryot Gene Expr* **16**, 211–31.

21. Parker, G. C., Anastassova-Kristeva, M., Eisenberg, L. M., Rao, M. S., Williams, M. A., Sanberg, P. R. and English, D. (2005). Stem cells: Shibboleths of development, Part 2: Toward a functional definition. *Stem Cells Dev* **14**, 463–9.

22. Spivakov, M. and Fisher, A. G. (2007). Epigenetic signatures of stem-cell identity. *Nat Rev Genet* **8**, 263–71.

23. Bibikova, M., Chudin, E., Wu, B., Zhou, L., Garcia, E. W., Liu, Y., Shin, S., Plaia, T. W., Auerbach, J. M., Arking, D. E., Gonzalez, R., Crook, J., Davidson, B., Schulz, T. C., Robins, A., Khanna, A., Sartipy, P., Hyllner, J., Vanguri, P., Savant-Bhonsale, S., Smith, A. K., Chakravarti, A., Maitra, A., Rao, M., Barker, D. L., Loring, J. F. and Fan, J. B. (2006). Human embryonic stem cells have a unique epigenetic signature. *Genome Res* **16**, 1075–83.

24. Miura, T., Luo, Y., Khrebtukova, I., Brandenberger, R., Zhou, D., Thies, R. S., Vasicek, T., Young, H., Lebkowski, J., Carpenter, M. K. and Rao, M. S. (2004). Monitoring early differentiation events in human embryonic stem cells by massively parallel signature sequencing and expressed sequence tag scan. *Stem Cells Dev* **13**, 694–715.

25. Takahashi, J. (2006). Stem cell therapy for Parkinson's disease. Ernst Schering Res Found Workshop 229–44.

26. Bonnevie, L., Bel, A., Sabbah, L., Al Attar, N., Pradeau, P., Weill, B., Le Deist, F., Bellamy, V., Peyrard, S., Menard, C., Desnos, M., Bruneval, P., Binder, P., Hagege, A. A., Puceat, M. and Menasche, P. (2007). Is xenotransplantation of embryonic stem cells a realistic option? *Transplantation* **83**, 333–5.

27. Hewitt, Z., Priddle, H., Thomson, A. J., Wojtacha, D. and McWhir, J. (2007). Ablation of undifferentiated human embryonic stem cells: Exploiting innate immunity against the gal alpha1–3galbeta1–4glcnac-r (alpha-gal) epitope. *Stem Cells* **25**, 10–18.

28. Ginis, I., Luo, Y., Miura, T., Thies, S., Brandenberger, R., Gerecht-Nir, S., Amit, M., Hoke, A., Carpenter, M. K., Itskovitz-Eldor, J. and Rao, M. S. (2004). Differences between human and mouse embryonic stem cells. *Dev Biol* **269**, 360–80.

29. Zhan, M., Miura, T., Xu, X. and Rao, M. S. (2005). Conservation and variation of gene regulation in embryonic stem cells assessed by comparative genomics. *Cell Biochem Biophys* **43**, 379–405.

30. Qazilbash, M. H., Saliba, R. M., Hosing, C., Mendoza, F., Qureshi, S. R., Weber, D. M., Wang, M., Flosser, T., Couriel, D. R., De Lima, M., Kebriaei, P., Popat, U., Alousi, A. M., Champlin, R. E. and Giralt, S. A. (2007). Autologous stem cell transplantation is safe and feasible in elderly patients with multiple myeloma. *Bone Marrow Transplant* **39**, 279–83.

31. Molcayni, M., Bentz, K., Maegele, M., Simanski, C., Carlitscheck, C., Schneider, A., Hescheler, J., Bouillon, B., Schafer, U. and Neugebauer, E. (2007). Embryonic stem cell transplantation after experimental traumatic brain injury dramatically improves neurological outcome, but may cause tumors. *J Neurotrauma* **24**, 216–25.

32. Cheng, A., Coksaygan, T., Tang, H., Khatri, R., Balice-Gordon, R. J., Rao, M. S. and Mattson, M. P. (2007). Truncated tyrosine kinase b brain-derived neurotrophic factor receptor directs cortical neural stem cells to a glial cell fate by a novel signaling mechanism. *J Neurochem* **100**, 1515–30.

Chapter 2

Adult Stem Cells for the Treatment of Neurological Disease

C.M. Rice and N.J. Scolding

Summary

The characteristic CNS responses to injury including increased cell production and attempts at regenerative repair – implicitly predicted where not directly demonstrated by Cajal, but only now more fully confirmed – have important implications for regenerative therapies. Spontaneous CNS cell replacement compares poorly with the regenerative functional repair seen elsewhere, but harnessing, stimulating or supplementing this process represents a new and attractive therapeutic concept.

Stem cells, traditionally defined as clone-forming, self-renewing, pluripotent progenitor cells, have already proved themselves to be an invaluable source of transplantation material in several clinical settings, most notably haematological malignancy, and attention is now turning to a wider variety of diseases in which there may be potential for therapeutic intervention with stem cell transplantation. Neurological diseases, with their reputation for relentless progression and incurability are particularly tantalising targets. The optimal source of stem cells remains to be determined but bone marrow stem cells may themselves be included amongst the contenders.

Any development of therapies using stem cells must depend on an underlying knowledge of their basic biology. The haemopoietic system has long been known to maintain circulating populations of cells with short life spans, and this system has greatly informed our knowledge of stem cell biology. In particular, it has helped yield the traditional stem cell model – a hierarchical paradigm of progressive lineage restriction. As cells differentiate, their fate choices become progressively more limited, and their capacity for proliferation reduced, until fully differentiated, mitotically quiescent cells are generated. Even this, however, is now under challenge.

Key words: Adult stem cells, Bone marrow, Neurodegenerative disease, Stem cell therapy

1. How Might Stem Cells Repair the Brain?

1.1. Traditional Stem Cell Paradigm

The haemopoietic stem cell (HSC) provides the template against which stem cells isolated from other tissues (including skin, liver, and pancreas) have been judged *(1, 2)*. The accepted model was

Neil J. Scolding and David Gordon (eds.), *Methods in Molecular Biology, Neural Cell Transplantation vol. 549*
DOI: 10.1007/978-1-60327-931-4_2, © Humana Press, a part of Springer Science+Business Media, LLC 2009

that stem cells were numerically insubstantial cells that divided asymmetrically to self-replicate (thereby maintaining a stem cell pool) and generate more committed precursors *(3, 4)*, which then could yield all the cell lineages present in their resident organ. Differentiation was therefore considered a unidirectional process of sequential, irreversible commitment steps, which progressively restricted available fate choices. A comparable process of progressive cell fate restriction was considered to equally apply to embryonic stem cells and their behaviour during organ and tissue development.

1.2. Challenges to the Traditional Hierarchical Stem Cell Model

The unexpected trans-differentiation potential of tissue-specific adult stem cells [reviewed in *(5–7)*] clearly challenges this traditional hierarchical model. Under conditions of strong selective pressure, intravenously delivered and highly purified HSCs restore biochemical function in murine tyrosinemia – a striking illustration of transplanted bone marrow-derived cells restoring function in an adult non-haemopoietic organ *(8)*. Human donor–recipient chimerism has been detected in the liver, brain, heart, and lungs of bone marrow transplant recipients, with apparently fully differentiated and functionally integrated bone marrow-derived cells in each organ *(9–13)*. Further studies of adult stem cells have also called into question the nature of differentiation from 'undifferentiated' stem cells, offering further challenges to the traditional linear model. The intuitively attractive hypothesis of a wholly undifferentiated cell coming to express more and more specialised markers of a particular lineage as it differentiates might not necessarily be the case. A number of studies have demonstrated the constitutive and simultaneous expression of markers of a number of cell lineages in non-embryonic human stem cells *(14)*, including, for example, that of 'mature' muscle protein *(15–17)*. Rather than exhibiting progressive lineage restriction, therefore, some types of stem cell may show from the outset molecular features of multi-differentiated cells, and multiple redundant gene expression could represent a cost of retaining multi-potentiality.

The implications of these often-surprising findings are potentially of considerable significance, of course, for our understanding of basic and developmental cell biology – but they also extend well beyond the laboratory. They raise the serious and initially unexpected possibility that adult stem cells might be of significant potential in developing cell-based therapies.

Initially, however, questions about the nature of adult, tissue-based stem cells were raised. In experimental disease studies, the degree of differentiation and integration of the transplanted adult cells often correlated poorly with the observed functional benefit – which indeed was in some instances reported in the absence of trans-differentiation *(18)*. The observation that bone marrow cells fused in vitro with embryonic stem cells *(19)* raised the possibility that fusion might mimic apparent adult cell trans-differentiation. Subsequently, two studies demonstrated the presence of genetic markers derived from both host and donor in individual liver cells following transplantation of bone marrow-derived cells *(20, 21)* – strongly suggesting fusion. Further studies using Cre/lox recombination also demonstrated fusion between bone marrow-derived cells and cardiomyocytes, hepatocytes, and Purkinje cells of the cerebellum *(22)*. However, the picture is complicated – no host-donor fusion was demonstrated in various other sites including renal glomeruli *(23)*, buccal epithelium *(24)*, pancreas *(25)*, and brain *(26)* following transplantation of adult bone marrow-derived cells.

There is a well-recognised association between aneuploidy (abnormal chromosomal number and arrangement) and neoplasia *(27, 28)*, but is fusion necessarily or always undesirable and unwanted? It is increasingly clear that polyploidy (abnormal chromosomal number) was previously under-recognised and can occur in the absence of malignancy – and in some circumstances may be advantageous. Aside from contributing to evolution and the development of new species *(29–31)*, polyploidy correlates with metabolic requirement and post-natal growth rate *(32, 33)* and occurs normally during insect embryogenesis *(34)*. Recently, a significant proportion of normal murine neural progenitors was found to be aneuploid *(35, 36)*.

In the adult mammal it is striking that organs with considerable regenerative capacity are frequently populated with cells that are not diploid; megakaryocytes are polyploid cells *(37)* and aneuploidy often occurs in the liver *(38–40)*. Polyploid cells have also been reported in the cerebellum *(41, 42)*, arterial smooth muscle *(43)*, heart *(44, 45)*, uterus *(46)*, and thyroid *(47, 48)* under normal circumstances or in response to stress. Marrow-derived donor cells have been identified in the liver, brain, and heart post-BMT *(49–53)*, and it is notable that these are the locations at which fusion has been most convincingly demonstrated *(54–56)*.

These findings raise the intriguing possibility that cell fusion may play crucial but hitherto unappreciated physiological roles in development, tissue maintenance, and repair *(57, 58)*. It seems likely that in vivo, under normal circumstances, diploid cells have a selective advantage over polyploid cells, but this balance may alter in disease states *(59)*. Certainly, cell fusion appears to be more common than previously recognised; polyploidy is not synonymous with disease, and whether fusion compromises or enhances the reparative capacity of stem cells remains to be determined.

Furthermore, this complex emerging picture stimulated a re-assessment first of the question of fusion, and whether it might not necessarily be a detrimental property vis-à-vis repair *(60)*, and consequently of the properties of adult stem cells, and how they might contribute to tissue repair.

1.3. Non-canonical Stem Cell Repair

Whilst there has been an increasing acceptance that adult stem cells do play a role in endogenous repair *(61)*, one of the most important developments in our understanding of adult stem cell biology in the last few years is that the ability to differentiate into different lineages is just one amongst a number of reparative properties they exhibit – and indeed it is now hypothesised by some that such 'non-canonical' stem cell activities such as stimulating angiogenesis, immune modulation, growth factor production, and protection against injury may be more important mechanistically in their regenerative capacity than what we have always assumed to be their core asset, multipotentiality *(60, 62–64)*.

This multiplicity and versatility of reparative functions has been studied particularly in bone marrow stem cells. We now know that stem cells can elaborate various growth factors *(65–71)* and, more than this, can protect axons in MS-related and other very different models of CNS damage *(69, 72–78)*. They have very pronounced immune modulating functions *(79–81)*; mesenchymal stem cells have been shown to affect both T and B cells *(82–84)* and their immunomodulatory effects have been employed clinically in the treatment of graft-versus-host disease *(85)*. Neural precursors from the rodent neonate *(86–88)* and adult *(89, 90)* also have immune suppressant effects. Adult stem cells additionally may exert significant paracrine effects on the differentiation of local endogenous neural stem cells *(91, 92)* and on other local cell populations and reparative processes, including angiogenesis *(93)*.

It should be emphasised that the involvement and importance of such 'non-canonical' mechanisms of repair in contributing to functional benefit has recently been confirmed similarly in relation to human neural stem cells, including those harvested de novo and those derived from embryonic stem cells *(94, 95)*.

2. Practical Aspects of Stem Cell Treatments

2.1. Safety

The proliferative potential of embryonic stem cells carries serious practical advantages – cells sufficient for numerous grafts might be prepared from a single sample – but is also responsible for perhaps the principal potential hazard, namely, the very real possibility of tumour formation *(96)*. Although some studies do suggest wide migration and appropriate differentiation of these cells without the development of teratomata in the neonatal brain *(97)*, studies in the more relevant environment of the adult brain, even using 'pre-differentiated' (dopaminergic) cells derived from embryonic stem cells continue to substantiate this risk *(98)*.

Perhaps the two most obvious and serious other adverse effects are immune rejection and infection – the latter applying particularly if stem cells are grown on non-human feeder cell layers, or if single, massively expanded cell lines are used to treat many patients *(99)*.

Neither of these potential dangers appears to apply in the specific case of autologous adult stem cell treatment. Donor-related transmissible infections plainly are not relevant to such an approach, while decades of delivering such cells to bone marrow transplant recipients has failed to provide evidence of donor-derived malignancy. This said, extensive in vitro proliferation or other manipulation of cells prior to infusion, whatever their source, may be presumed to carry a greater risk of tumourigenesis than the use of untreated cells, though such a risk has been shown more convincingly for embryonic stem cells *(98, 100, 101)* than for adult. Further witness to the safety of adult bone marrow stem cell transplantation is offered by the rapidly increasing experience of cardiologists in using bone marrow cells to aid cardiac repair *(102–104)*, and the longer term safety data emerging in other fields of (mostly oncological) medicine *(105)*, in the treatment of graft-versus-host disease *(106)* and the enhancement of engraftment after bone marrow transplantation *(107)*.

Finally, in the case of treating CNS disease, concerns regarding the potential for cells to cause epileptogenic foci *(108)* should also be mentioned.

2.2. Delivery

How best to get cells to the place of disease plainly represents a very serious question in the case of neurodegenerative disease. Direct intra-parenchymal injection, much as originally executed for foetal transplants in Parkinson's disease *(109, 110)* (PD) and more recently in the case of growth factor treatments for PD *(111)* is technically demanding and potentially hazardous, especially if, as would be the case for some diseases, multiple injections were necessary. Experimental studies have, however, offered the attractive alternative possibility that many stem cell types might find access to the CNS following intravenous delivery and indeed appear capable of homing to sites of disease within the brain, as well as migrating and surviving within the adult CNS *(112–114)*.

Powerful evidence supporting this approach comes from a number of postmortem studies of individuals who have been treated for leukaemic (or other usually haematological conditions) using bone marrow transplantation, and who, more specifically, have received marrow from donors who are HLA-matched but gender mismatched. In these individuals, donor cells can readily be identified indefinitely after the procedure by sex chromosome labelling. In such individuals, the demonstration years or even decades after transplantation of small numbers of donor-derived cells in a number of tissues, most notably from a neurological perspective including brain *(115, 116)* and muscle *(117)*, often of highly differentiated morphology appropriate to their environment (for example, Purkinje cells in the cerebellum) and apparently fully integrated into that tissue, offers persuasive evidence of the lasting functional pluripotentiality of intravenously delivered bone marrow-derived stem cells.

2.3. Accessibility

The ease of collection of bone marrow cells either by cytokine-driven peripheral collection or by bone marrow harvest under a short general anaesthetic makes bone marrow an attractive, ethically robust source of stem cells. Depending on the nature of the disease being treated, the potential for the use of autologous cells is an additional advantage. Alternatively, allogenic transplantation or genetic modification might be more appropriate for diseases such as inborn errors of metabolism. As previously mentioned however, this would intuitively be associated with a higher risk.

3. Stem Cells in Specific Neurological Diseases

By way of illustrating progress in this fast-changing area, it might be helpful to mention progress in the development of stem cell treatments of some relatively common neurological conditions. It should be emphasised that the list that follows is by no means exhaustive, but is intended to illustrate that the methodological and technical chapters that follow have not just neurobiological relevance, but may also be set in a clinical neurological context.

3.1. Parkinson's Disease

Parkinson's disease might reasonably claim *alpha* and *omega* positions in neurological cell therapy! It was the first neurodegenerative disease treated surgically with cell implantation – and, around two decades later and largely in reaction to the side effects reported in two large-scale controlled therapeutic trials of embryonic cell therapy, the first to be subject to an internationally agreed moratorium on further similar studies. None were studies, it must be emphasised, using stem cells.

The attraction of idiopathic PD as a disease appropriate for such intervention centres on the focal nature of degeneration of just one class of neurone, the dopaminergic cell, in the *substantia nigra*, at least (it is believed) in early disease. Moreover, faithful reconstruction of the nigrostriatal pathway does not seem necessary for clinically significant benefit.

Experimental studies using embryonic and adult stem cells have, of course, continued and both show some promise *(118–120)*. However, the potential for tumour formation when ES cells are used has been well-illustrated; means of reducing the possibility have been explored, but dopaminergic pre-differentiation and purification by no means abolish this risk *(98, 121)*.

3.2. Huntington's Disease

Huntington's disease (HD) is an autosomal dominant condition causing psychiatric or neurological symptoms and, as the disease progresses, chorea and dementia. Gaba-ergic medium spiny neurones are lost in the striatum, and experimental attempts to treat HD models using implanted foetal striatal neurones commenced over two decades ago [reviewed in *(122)*], many achieving considerable success. Phase 1 clinical trials followed *(123)* with the hope that clinical benefit might include improvement in both movement disorder and cognitive function *(124)*.

Alternative stem cell sources have also been investigated: adult brain-derived and bone marrow cells can elicit functional benefit in rodent models *(125, 126)* – and again in the case of bone marrow cells, the benefit was suggested to be more through growth factor elaboration than trans-differentiation *(125)*.

3.3. Multiple Sclerosis

In multiple sclerosis, damage to oligodendrocytes and their myelin sheaths causes demyelination in lesions throughout the CNS *(127)*. Axons are also damaged acutely *(128, 129)*, irreversible axon loss assuming increasing functional relevance with disease duration *(130, 131)*. Spontaneous myelin repair occurs *(132–136)* and contributes to functional recovery *(137)*; there is evidence that it may also help preserve axonal integrity *(130, 138)*. It is, however, of limited ultimate success. When repair fails, permanent demyelination and accumulating axon loss cause increasing irreversible disability. Successful remyelination in rodents after implantation of oligodendrocyte lineage cells *(139–146)* or various cell lines *(142, 147, 148)*, with resulting improved conduction *(149)* and function *(150)*, has provided powerful and optimistic stimuli for the exploration of therapeutic remyelination strategies for patients.

Of the principal hurdles to clinical studies, addressing the multi-focal nature of the disease and, of course, sourcing cells are not the least important. Adult stem cells have some attractions in both respects. Adult neural stem cells ameliorate experimental models of MS *(86, 89)*, promoting multifocal remyelination and functional recovery after intravenous or intrathecal injection *(89)*. Cells of bone marrow origin can also promote experimental remyelination and recovery following both focal implantation and intravenous infusion *(151–154)*. Interestingly, in the case of both cell types, neuroprotective effects in reducing axon loss *(75)*, and immune suppression also *(79, 80, 87, 88, 90)*, have been shown to represent initially unexpected beneficial effects of the infused stem cell, effects which are considered by many to be more important reparatively than their trans-differentiation potential. Clinical studies exploring the safety and feasibility of autologous bone marrow stem cell treatment (without myeloablative pre-conditioning) have now been commenced *(155)*.

3.4. Amyotrophic Lateral Sclerosis

Amyotrophic lateral sclerosis (ALS) is a progressive condition, which results in widespread muscle denervation due to the loss of both central and peripheral motor neurons. As cells possessing amongst the longest axons in the body – from motor cortex to synapse in the sacral spinal cord segments – they are not readily amenable to cellular replacement using stem cells or other potential sources of motor neurones.

Nonetheless, haematogenous and intrathecal delivery of stem cells has been considered, studied in experimental models, and even used in clinical trials with peripheral blood and with bone marrow-derived stem cells *(156, 157)*. Again, the non-canonical reparative properties of such cells may be of no less importance than neural differentiation and cell replacement. Local neuroprotection may be particularly valuable, but an additional possibility has emerged from recent studies of transgeneic 'ALS' mice. Here, evidence of increased stem cell proliferation in the spinal cord of affected adult mice has reasonably been interpreted as indicating an abortive regenerative or reparative response *(158)*. The ability of BM stem cells to enhance local reparative mechanisms might therefore also be of potential relevance in this clinical setting.

3.5. Stroke

Arguably, the potential benefits of adult stem cell therapy in stroke might, as with ALS, depend more on the alternative pathways of stimulating and promoting repair than on trans-differentiation. The problems of restoring connectivity over potentially long distances are not dissimilar – though the multiplicity of tracts and connections disrupted in stroke is potentially larger and broader.

However, a single ischaemic insult to the brain is commonly followed by significant functional recovery, and conventionally, this has been explained by a combination of factors, including neuronal plasticity, and a reversible component of the vascular injury, perhaps particularly demyelination, in the ischaemic penumbra. Increasingly it is accepted that other endogenous repair mechanisms are triggered and contribute, and these could potentially be enhanced by stem cell therapies.

Certainly, experimental studies have suggested that bone marrow cells are recruited to ischaemic brain *(159)*, and that they may enhance functional improvement in animal models of stroke when injected focally *(160, 161)* or delivered intravenously *(162)* – though the mechanisms have not fully been elucidated.

New vessel formation, improving perfusion, represents one aspect of the endogenous repair response, and this does appear to be enhanced by transplanted bone marrow stem cells of various types *(163–165)*. Neural plasticity itself is also increased *(164)*. Endogenous neurogenesis is increased in response to ischaemia *(164, 166)*; it can be enhanced by extraneous factors [including environmental enrichment *(167)*]. Exogenous (or endogenously migrating) bone marrow cells may also increase such neurogenesis *(168)*, perhaps through growth factor elaboration *(164, 168)*.

As is the case with MS and its models, early pilot translational studies in patients have begun even while experimental analysis and exploration of the mechanisms of potential benefit are continuing in various disease models. Several years ago, a study using a human neuronal cell line implanted into patients was reported *(169)*, and more recently a pilot study of intravenously delivered bone marrow stem cells was described *(170)*. Finally, a possible insight into the future has been offered by a clinical study exploring putative CNS regenerative treatment not by autologous bone marrow stem cell transplantation or infusion, but therapeutic mobilisation of bone marrow stem cells by intravenously delivered granulocyte-colony-stimulating factor in patients with stroke *(171)*.

References

1. Dexter, T. M., Ponting, I. L., Roberts, R. A., Spooncer, E., Heyworth, C. & Gallagher, J. T. (1988). Growth and differentiation of hematopoietic stem cells. *Soc. Gen. Physiol Ser.* **43**: 25–38.

2. Martin-Rendon, E. & Watt, S. M. (2003). Stem cell plasticity. *Br. J. Haematol.* **122**: 877–891.

3. Till, J. E. & McCulloch, E. A. (1961). A direct measurement of the radiation sensitivity of normal mouse bone marrow cells. *Radiat. Res.* **14**: 213–222.

4. Weissman, I. L. (2000). Stem cells: units of development, units of regeneration, and units in evolution. *Cell* **100**: 157–168.

5. Martin-Rendon, E. & Watt, S. M. (2003). Exploitation of stem cell plasticity. *Transfus. Med.* **13**: 325–349.

6. Morrison, S. J. (2001). Stem cell potential: can anything make anything? Curr. *Biol.* **11**: R7–R9.

7. Weissman, I. L., Anderson, D. J. & Gage, F. (2001). Stem and progenitor cells: origins, phenotypes, lineage commitments, and transdifferentiations. *Annu. Rev. Cell Dev. Biol* **17**: 387–403.

8. Lagasse, E., Connors, H., Al Dhalimy, M., Reitsma, M., Dohse, M., Osborne, L., Wang, X., Finegold, M., Weissman, I. L. & Grompe, M. (2000). Purified hematopoietic stem cells can differentiate into hepatocytes in vivo. *Nat. Med.* **6**: 1229–1234.

9. Alison, M. R., Poulsom, R., Jeffery, R., Dhillon, A. P., Quaglia, A., Jacob, J., Novelli, M., Prentice, G., Williamson, J. & Wright, N. A. (2000). Hepatocytes from non-hepatic adult stem cells. *Nature* **406**: 257.

10. Suratt, B. T., Cool, C. D., Serls, A. E., Chen, L., Varella-Garcia, M., Shpall, E. J., Brown, K. K. & Worthen, G. S. (2003). Human pulmonary chimerism after hematopoietic stem cell transplantation. *Am. J. Respir. Crit Care Med.* **168**: 318–322.

11. Theise, N. D., Nimmakayalu, M., Gardner, R., Illei, P. B., Morgan, G., Teperman, L., Henegariu, O. & Krause, D. S. (2000). Liver from bone marrow in humans. *Hepatology* **32**: 11–16.

12. Thiele, J., Varus, E., Wickenhauser, C., Kvasnicka, H. M., Metz, K. A. & Beelen, D. W. (2004). Regeneration of heart muscle tissue:

quantification of chimeric cardiomyocytes and endothelial cells following transplantation. *Histol. Histopathol.* **19**: 201–209.

13. Weimann, J. M., Charlton, C. A., Brazelton, T. R., Hackman, R. C. & Blau, H. M. (2003). Contribution of transplanted bone marrow cells to Purkinje neurons in human adult brains. *Proc. Natl. Acad. Sci. USA* **100**: 2088–2093.

14. Ji, K. H., Xiong, J., Hu, K. M., Fan, L. X., Liu, H. Q. (2008). Simultaneous expression of Oct4 and genes of three germ layers in single cell-derived multipotent adult progenitor cells. *Ann. Hematol.* **87**: 431–438.

15. Arsic, N., Zacchigna, S., Zentilin, L., Ramirez-Correa, G., Pattarini, L., Salvi, A., Sinagra, G. & Giacca, M. (2004). Vascular endothelial growth factor stimulates skeletal muscle regeneration in vivo. *Mol. Ther.* **10**: 844–854.

16. Burattini, S., Ferri, P., Battistelli, M., Curci, R., Luchetti, F. & Falcieri, E. (2004). C2C12 murine myoblasts as a model of skeletal muscle development: morpho-functional characterization. *Eur. J. Histochem.* **48**: 223–233.

17. Bossolasco, P., Corti, S., Strazzer, S., Borsotti, C., Del, B. R., Fortunato, F., Salani, S., Quirici, N., Bertolini, F. et al. (2004). Skeletal muscle differentiation potential of human adult bone marrow cells. *Exp. Cell Res.* **295**: 66–78.

18. Balsam, L. B., Wagers, A. J., Christensen, J. L., Kofidis, T., Weissman, I. L. & Robbins, R. C. (2004). Haematopoietic stem cells adopt mature haematopoietic fates in ischaemic myocardium. *Nature* **428**: 668–673.

19. Terada, N., Hamazaki, T., Oka, M., Hoki, M., Mastalerz, D. M., Nakano, Y., Meyer, E. M., Morel, L., Petersen, B. E. & Scott, E. W. (2002). Bone marrow cells adopt the phenotype of other cells by spontaneous cell fusion. *Nature* **416**: 542–545.

20. Vassilopoulos, G., Wang, P. R. & Russell, D. W. (2003). Transplanted bone marrow regenerates liver by cell fusion. *Nature* **422**: 901–904.

21. Wang, X., Willenbring, H., Akkari, Y., Torimaru, Y., Foster, M., Al Dhalimy, M., Lagasse, E., Finegold, M., Olson, S. & Grompe, M. (2003). Cell fusion is the principal source of bone-marrow-derived hepatocytes. *Nature* **422**: 897–901.

22. Alvarez-Dolado, M., Pardal, R., Garcia-Verdugo, J. M., Fike, J. R., Lee, H. O., Pfeffer, K., Lois, C., Morrison, S. J. & Alvarez-Buylla, A. (2003). Fusion of bone-marrow-derived cells with Purkinje neurons, cardiomyocytes and hepatocytes. *Nature* **425**: 968–973.

23. Masuya, M., Drake, C. J., Fleming, P. A., Reilly, C. M., Zeng, H., Hill, W. D., Martin-Studdard, A., Hess, D. C. & Ogawa, M. (2003). Hematopoietic origin of glomerular mesangial cells. *Blood* **101**: 2215–2218.

24. Tran, S. D., Pillemer, S. R., Dutra, A., Barrett, A. J., Brownstein, M. J., Key, S., Pak, E., Leakan, R. A., Kingman, A. et al. (2003). Differentiation of human bone marrow-derived cells into buccal epithelial cells in vivo: a molecular analytical study. *Lancet* **361**: 1084–1088.

25. Ianus, A., Holz, G. G., Theise, N. D. & Hussain, M. A. (2003). In vivo derivation of glucose-competent pancreatic endocrine cells from bone marrow without evidence of cell fusion. *J. Clin. Invest.* **111**: 843–850.

26. Corti, S., Locatelli, F., Strazzer, S., Salani, S., Del Bo, R., Soligo, D., Bossolasco, P., Bresolin, N., Scarlato, G. & Comi, G. P. (2002). Modulated generation of neuronal cells from bone marrow by expansion and mobilization of circulating stem cells with in vivo cytokine treatment. *Exp. Neurol.* **177**: 443–452.

27. Barlogie, B. (1984) Abnormal cellular DNA content as a marker of neoplasia. *Eur. J. Cancer Clin. Oncol.* **20**: 1123–1125.

28. Storchova, Z. & Pellman, D. (2004) From polyploidy to aneuploidy, genome instability and cancer. *Nat. Rev. Mol. Cell Biol* **5**: 45–54.

29. Otto, S. P. & Whitton, J. (2000). Polyploid incidence and evolution. *Annu. Rev. Genet.* **34**: 401–437.

30. Soltis, D. E. & Soltis, P. S. (1999). Polyploidy: recurrent formation and genome evolution. *Trends Ecol. Evol.* **14**: 348–352.

31. Bailey, J. A., Gu, Z., Clark, R. A., Reinert, K., Samonte, R. V., Schwartz, S., Adams, M. D., Myers, E. W., Li, P. W. & Eichler, E. E. (2002). Recent segmental duplications in the human genome. *Science* **297**: 1003–1007.

32. Brodsky, V. Y. & Delone, G. V. (1990). Functional control of hepatocyte proliferation. Comparison with the temporal control of cardiomyocyte proliferation. *Biomed. Sci.* **1**: 467–470.

33. Vinogradov, A. E., Anatskaya, O. V. & Kudryavtsev, B. N. (2001). Relationship of hepatocyte ploidy levels with body size and growth rate in mammals. *Genome* **44**: 350–360.

34. Edgar, B. A. & Orr-Weaver, T. L. (2001). Endoreplication cell cycles: more for less. *Cell* **105**: 297–306.

35. Kaushal, D., Contos, J. J., Treuner, K., Yang, A. H., Kingsbury, M. A., Rehen, S. K., McConnell, M. J., Okabe, M., Barlow, C. & Chun, J. (2003). Alteration of gene expression by chromosome loss in the postnatal mouse brain. *J Neurosci.* **23**: 5599–5606.

36. Rehen, S. K., McConnell, M. J., Kaushal, D., Kingsbury, M. A., Yang, A. H. & Chun, J. (2001). Chromosomal variation in neurons of the developing and adult mammalian nervous system. *Proc. Natl. Acad. Sci. USA* **98**: 13361–13366.

37. Ravid, K., Lu, J., Zimmet, J. M. & Jones, M. R. (2002). Roads to polyploidy: the megakaryocyte example. *J Cell Physiol* **190**: 7–20.

38. Arias I.M., Boyer, J.L., Fausto, N., Jakoby, W.B., Schachter, D. A. & Shafritz, D. A. (2001). The Liver: Biology and Pathobiology. Lippincott Williams and Wilkins, Philadelphia.

39. Brodsky, W. Y. & Uryvaeva, I. V. (1977). Cell polyploidy: its relation to tissue growth and function. *Int. Rev. Cytol.* **50**: 275–332.

40. Kudryavtsev, B. N., Kudryavtseva, M. V., Sakuta, G. A. & Stein, G. I. (1993). Human hepatocyte polyploidization kinetics in the course of life cycle. *Virchows Arch. B Cell Pathol. Incl. Mol. Pathol.* **64**: 387–393.

41. Lapham, L. W. (1968). Tetraploid DNA content of Purkinje neurons of human cerebellar cortex. *Science* **159**: 310–312.

42. Mares, V., Lodin, Z. & Sacha, J. (1973). A cytochemical and autoradiographic study of nuclear DNA in mouse Purkinje cells. *Brain Res.* **53**: 273–289.

43. Barrett, T. B., Sampson, P., Owens, G. K., Schwartz, S. M. & Benditt, E. P. (1983). Polyploid nuclei in human artery wall smooth muscle cells. *Proc. Natl. Acad. Sci. USA* **80**: 882–885.

44. Piper, H. M. & Isenberg, G. (1989). Isolated Adult Cardiomyocytes: Structure and Metabolism. CRC Press, Boca Raton.

45. Sandritter, W. & Scomazzoni, G. (1964). Deoxyribonucleic acid content (feulgen photometry) and dry weight (interference microscopy) of normal and hypertrophic heart muscle fibers. *Nature* **202**: 100–101.

46. Heiden, F. L. & James, J. (1975). Polyploidy in the human myometrium. *Z. Mikrosk Anat Forsch* **89**: 18–26.

47. Auer, G. U., Backdahl, M., Forsslund, G. M. & Askensten, U. G. (1985). Ploidy levels in nonneoplastic and neoplastic thyroid cells. *Anal. Quant. Cytol. Histol.* 7: 97–106.

48. Gilbert, P. & Pfitzer, P. (1977). Facultative polyploidy in endocrine tissues. *Virchows Arch. B Cell Pathol.* **25**: 233–242.

49. Alison, M. R., Poulsom, R., Jeffery, R., Dhillon, A.P., Quaglia, A., Jacob, J., Novelli, M., Prentice, G., Williamson, J. & Wright, N. A. (2000). Hepatocytes from non-hepatic adult stem cells. *Nature* **406**: 257.

50. Mezey, E., Key, S., Vogelsang, G., Szalayova, I., Lange, G. D. & Crain, B. (2003). Transplanted bone marrow generates new neurons in human brains. *Proc. Natl. Acad. Sci. USA* **100**: 1364–1369.

51. Theise, N. D., Nimmakayalu, M., Gardner, R., Illei, P. B., Morgan, G., Teperman, L., Henegariu, O. & Krause, D. S. (2000). Liver from bone marrow in humans. *Hepatology* **32**: 11–16.

52. Thiele, J., Varus, E., Wickenhauser, C., Kvasnicka, H. M., Metz, K. A. & Beelen, D. W. (2004). Regeneration of heart muscle tissue: quantification of chimeric cardiomyocytes and endothelial cells following transplantation. *Histol. Histopathol.* **19**: 201–209.

53. Weimann, J. M., Charlton, C. A., Brazelton, T. R., Hackman, R. C. & Blau, H. M. (2003) Contribution of transplanted bone marrow cells to Purkinje neurons in human adult brains. *Proc. Natl. Acad. Sci. USA* **100**: 2088–2093.

54. Alvarez-Dolado, M., Pardal, R., Garcia-Verdugo, J. M., Fike, J. R., Lee, H. O., Pfeffer, K., Lois, C., Morrison, S. J. & Alvarez-Buylla, A. (2003). Fusion of bone-marrow-derived cells with Purkinje neurons, cardiomyocytes and hepatocytes. *Nature* **425**: 968–973.

55. Vassilopoulos, G., Wang, P. R. & Russell, D. W. (2003) Transplanted bone marrow regenerates liver by cell fusion. *Nature* **422**: 901–904.

56. Wang, X., Willenbring, H., Akkari, Y., Torimaru, Y., Foster, M., Al Dhalimy, M., Lagasse, E., Finegold, M., Olson, S. & Grompe, M. (2003). Cell fusion is the principal source of bone-marrow-derived hepatocytes. *Nature* **422**: 897–901.

57. Blau, H. M. (2002). A twist of fate. *Nature* **419**: 437.

58. Korbling, M. & Estrov, Z. (2003). Adult stem cells for tissue repair – a new therapeutic concept? *N. Engl. J. Med.* **349**: 570–582.

59. McKay, R. (2002). A more astonishing hypothesis. *Nat. Biotechnol.* **20**: 426–427.

60. Rice, C. M. & Scolding, N. J. (2004). Adult stem cells – reprogramming neurological repair? *Lancet* **364**: 193–199.

61. Korbling, M. & Estrov, Z. (2003). Adult stem cells for tissue repair. *N. Engl. J. Med* **349**: 570–582.

62. Quesenberry, P. J., Dooner, G., Dooner, M. & Abedi, M. (2005). Developmental biology: Ignoratio elenchi: red herrings in stem cell research. *Science* **308**: 1121–1122.

63. Miller, R. H. & Bai, L. (2007). The expanding influence of stem cells in neural repair. *Ann. Neurol.* **61**: 187–188.

64. Erlandsson, A. & Morshead, C. M. (2006). Exploiting the properties of adult stem cells for the treatment of disease. *Curr. Opin. Mol. Ther.* **8**: 331–337.

65. Guo, X., Du, J., Zheng, Q., Yang, S., Liu, Y., Duan, D. & Yi, C. (2002). Expression of transforming growth factor beta 1 in mesenchymal stem cells: potential utility in molecular tissue engineering for osteochondral repair. *J. Huazhong. Univ. Sci. Technol. Med. Sci.* 22: 112–115.

66. Kim, D. H., Yoo, K. H., Choi, K. S., Choi, J., Choi, S. Y., Yang, S. E., Yang, Y. S., Im, H. J., Kim, K. H. et al. (2005). Gene expression profile of cytokine and growth factor during differentiation of bone marrow-derived mesenchymal stem cell. *Cytokine* 31: 119–126.

67. Mayer, H., Bertram, H., Lindenmaier, W., Korff, T., Weber, H. & Weich, H. (2005). Vascular endothelial growth factor (VEGF-A) expression in human mesenchymal stem cells: autocrine and paracrine role on osteoblastic and endothelial differentiation. *J. Cell Biochem.* 95: 827–839.

68. Arnhold, S., Klein, H., Klinz, F. J., Absenger, Y., Schmidt, A., Schinkothe, T., Brixius, K., Kozlowski, J., Desai, B. et al. (2006). Human bone marrow stroma cells display certain neural characteristics and integrate in the subventricular compartment after injection into the liquor system. *Eur. J. Cell Biol.* 85: 551–565.

69. Chen, Q., Long, Y., Yuan, X., Zou, L., Sun, J., Chen, S., Perez-Polo, J. R. & Yang, K. (2005). Protective effects of bone marrow stromal cell transplantation in injured rodent brain: synthesis of neurotrophic factors. *J. Neurosci. Res.* 80: 611–619.

70. Ye, M., Chen, S., Wang, X., Qi, C., Lu, G., Liang, L. & Xu, J. (2005). Glial cell line-derived neurotrophic factor in bone marrow stromal cells of rat. *Neuroreport* 16: 581–584.

71. Garcia, R., Aguiar, J., Alberti, E., de la Cuetara, K. & Pavon, N. (2004). Bone marrow stromal cells produce nerve growth factor and glial cell line-derived neurotrophic factors. *Biochem. Biophys. Res. Commun.* 316: 753–754.

72. Madhavan, L., Ourednik, V. & Ourednik, J. (2005). Grafted neural stem cells shield the host environment from oxidative stress. *Ann. N Y Acad. Sci.* 1049: 185–188.

73. Ryu, J. K., Kim, J., Cho, S. J., Hatori, K., Nagai, A., Choi, H. B., Lee, M. C., McLarnon, J. G. & Kim, S. U. (2004). Proactive transplantation of human neural stem cells prevents degeneration of striatal neurons in a rat model of Huntington disease. *Neurobiol. Dis.* 16: 68–77.

74. Yasuhara, T., Matsukawa, N., Hara, K., Yu, G., Xu, L., Maki, M., Kim, S. U. & Borlongan, C. V. (2006). Transplantation of human neural stem cells exerts neuroprotection in a rat model of Parkinson's disease. *J. Neurosci.* 26: 12497–12511.

75. Zhang, J., Li, Y., Lu, M., Cui, Y., Chen, J., Noffsinger, L., Elias, S. B. & Chopp, M. (2006). Bone marrow stromal cells reduce axonal loss in experimental autoimmune encephalomyelitis mice. *J. Neurosci. Res.* 84: 587–595.

76. Simard, A. R. & Rivest, S. (2006). Neuroprotective properties of the innate immune system and bone marrow stem cells in Alzheimer's disease. *Mol. Psychiatry* 11: 327–335.

77. Isele, N. B., Lee, H. S., Landshamer, S., Straube, A., Padovan, C. S., Plesnila, N. & Culmsee, C. (2007). Bone marrow stromal cells mediate protection through stimulation of PI3-K/Akt and MAPK signaling in neurons. *Neurochem. Int.* 50: 243–250.

78. Priller, J., Persons, D. A., Klett, F. F., Kempermann, G., Kreutzberg, G. W. & Dirnagl, U. (2001). Neogenesis of cerebellar Purkinje neurons from gene-marked bone marrow cells in vivo. *J. Cell Biol.* 155: 733–738.

79. Zappia, E., Casazza, S., Pedemonte, E., Benvenuto, F., Bonanni, I., Gerdoni, E., Giunti, D., Ceravolo, A., Cazzanti, F. et al. (2005). Mesenchymal stem cells ameliorate experimental autoimmune encephalomyelitis inducing T-cell anergy. *Blood* 106: 1755–1761.

80. Uccelli, A., Pistoia, V. & Moretta, L. (2007). Mesenchymal stem cells: a new strategy for immunosuppression? *Trends Immunol.* 28: 219–226.

81. Stagg, J. (2007). Immune regulation by mesenchymal stem cells: two sides to the coin. *Tissue Antigens* 69: 1–9.

82. Corcione, A., Benvenuto, F., Ferretti, E., Giunti, D., Cappiello, V., Cazzanti, F., Risso, M., Gualandi, F., Mancardi, G. L. et al. (2006). Human mesenchymal stem cells modulate B-cell functions. *Blood* 107: 367–372.

83. Zappia, E., Casazza, S., Pedemonte, E., Benvenuto, F., Bonanni, I., Gerdoni, E., Giunti, D., Ceravolo, A., Cazzanti, F. et al. (2005). Mesenchymal stem cells ameliorate experimental autoimmune encephalomyelitis inducing T-cell anergy. *Blood* 106: 1755–1761.

84. Le, B. K., Tammik, L., Sundberg, B., Haynesworth, S. E. & Ringden, O. (2003). Mesenchymal stem cells inhibit and stimulate mixed lymphocyte cultures and mitogenic responses independently of the major histocompatibility complex. *Scand. J. Immunol.* 57: 11–20.

85. Ringden, O., Uzunel, M., Rasmusson, I., Remberger, M., Sundberg, B., Lonnies, H., Marschall, H. U., Dlugosz, A., Szakos, A. et al. (2006). Mesenchymal stem cells for treatment of therapy-resistant graft-versus-host disease. *Transplantation* 81: 1390–1397.

86. Ben-Hur, T., Einstein, O., Mizrachi-Kol, R., Ben-Menachem, O., Reinhartz, E., Karussis, D. & Abramsky, O. (2003). Transplanted

multipotential neural precursor cells migrate into the inflamed white matter in response to experimental autoimmune encephalomyelitis. *Glia* **41**: 73–80.

87. Einstein, O., Fainstein, N., Vaknin, I., Mizrachi-Kol, R., Reihartz, E., Grigoriadis, N., Lavon, I., Baniyash, M., Lassmann, H. & Ben-Hur, T. (2007) .Neural precursors attenuate autoimmune encephalomyelitis by peripheral immunosuppression. *Ann. Neurol* **61**: 209–218.

88. Einstein, O., Grigoriadis, N., Mizrachi-Kol, R., Reinhartz, E., Polyzoidou, E., Lavon, I., Milonas, I., Karussis, D., Abramsky, O. & Ben-Hur, T. (2006). Transplanted neural precursor cells reduce brain inflammation to attenuate chronic experimental autoimmune encephalomyelitis. *Exp. Neurol.* **198**: 275–284.

89. Pluchino, S., Quattrini, A., Brambilla, E., Gritti, A., Salani, G., Dina, G., Galli, R., Del Carro, U., Amadio, S. et al. (2003). Injection of adult neurospheres induces recovery in a chronic model of multiple sclerosis. *Nature* **422**: 688–694.

90. Pluchino, S., Zanotti, L., Rossi, B., Brambilla, E., Ottoboni, L., Salani, G., Martinello, M., Cattalini, A., Bergami, A. et al. (2005). Neurosphere-derived multipotent precursors promote neuroprotection by an immunomodulatory mechanism. *Nature* **436**: 266–271.

91. Rivera, F. J., Couillard-Despres, S., Pedre, X., Ploetz, S., Caioni, M., Lois, C., Bogdahn, U. & Aigner, L. (2006). Mesenchymal stem cells instruct oligodendrogenic fate decision on adult neural stem cells. *Stem Cells* **24**: 2209–2219.

92. Bai, L., Caplan, A., Lennon, D. & Miller, R. H. (2007). Human mesenchymal stem cells signals regulate neural stem cell fate. Neurochem. Res. 32: 353–362.

93. Kinnaird, T., Stabile, E., Burnett, M. S., Lee, C. W., Barr, S., Fuchs, S. & Epstein, S. E. (2004). Marrow-derived stromal cells express genes encoding a broad spectrum of arteriogenic cytokines and promote in vitro and in vivo arteriogenesis through paracrine mechanisms. Circ. Res. 94:678.

94. Redmond, D. E., Jr., Bjugstad, K. B., Teng, Y. D., Ourednik, V., Ourednik, J., Wakeman, D. R., Parsons, X. H., Gonzalez, R., Blanchard, B. C. et al. (2007). From the cover: behavioral improvement in a primate Parkinson's model is associated with multiple homeostatic effects of human neural stem cells. *Proc. Natl. Acad. Sci. USA* **104**: 12175–12180.

95. Lee, J. P., Jeyakumar, M., Gonzalez, R., Takahashi, H., Lee, P. J., Baek, R. C., Clark, D.,

Rose, H., Fu, G. et al. (2007). Stem cells act through multiple mechanisms to benefit mice with neurodegenerative metabolic disease. Nat. Med. 13: 439–447.

96. Reya, T., Morrison, S. J., Clarke, M. F. & Weissman, I. L. (2001). Stem cells, cancer, and cancer stem cells. Nature 414: 105–111.

97. Zhang, S. C., Wernig, M., Duncan, I. D., Brustle, O. & Thomson, J. A. (2001). In vitro differentiation of transplantable neural precursors from human embryonic stem cells. Nat. Biotechnol. 19: 1129–1133.

98. Thinyane, K., Baier, P. C., Schindehutte, J., Mansouri, A., Paulus, W., Trenkwalder, C., Flugge, G., Fuchs, E. (2005). Fate of pre-differentiated mouse embryonic stem cells transplanted in unilaterally 6-hydroxydopamine lesioned rats: histological characterization of the grafted cells. *Brain Res.* **1045**: 80–87.

99. Braude, P., Minger, S. L. & Warwick, R. M. (2005). Stem cell therapy: hope or hype? *BMJ* **330**: 1159–1160.

100. Draper, J. S., Smith, K., Gokhale, P., Moore, H. D., Maltby, E., Johnson, J., Meisner, L., Zwaka, T. P., Thomson, J. A. & Andrews, P. W. (2004). Recurrent gain of chromosomes 17q and 12 in cultured human embryonic stem cells. *Nat. Biotechnol.* **22**: 53–54.

101. Maitra, A., Arking, D. E., Shivapurkar, N., Ikeda, M., Stastny, V., Kassauei, K., Sui, G., Cutler, D. J., Liu, Y. et al. (2005). Genomic alterations in cultured human embryonic stem cells. *Nat Genet* **37**: 1099–1103.

102. Assmus, B., Honold, J., Schachinger, V., Britten, M. B., Fischer-Rasokat, U., Lehmann, R., Teupe, C., Pistorius, K., Martin, H. et al. (2006). Transcoronary transplantation of progenitor cells after myocardial infarction. *N. Engl. J. Med.* **355**: 1222–1232.

103. Lunde, K., Solheim, S., Aakhus, S., Arnesen, H., Abdelnoor, M., Egeland, T., Endresen, K., Ilebekk, A., Mangschau, A. et al. (2006). Intracoronary injection of mononuclear bone marrow cells in acute myocardial infarction. *N. Engl. J. Med.* **355**: 1199–1209.

104. Schachinger, V., Erbs, S., Elsasser, A., Haberbosch, W., Hambrecht, R., Holschermann, H., Yu, J., Corti, R., Mathey, D. G. et al. (2006). Intracoronary bone marrow-derived progenitor cells in acute myocardial infarction. *N. Engl. J. Med.* **355**: 1210–1221.

105. Koc, O. N., Gerson, S. L., Cooper, B. W., Dyhouse, S. M., Haynesworth, S. E., Caplan, A. I. & Lazarus, H. M. (2000). Rapid hematopoietic recovery after coinfusion of autologous-blood stem cells and culture-expanded marrow mesenchymal stem cells in advanced breast cancer patients receiving

high-dose chemotherapy. *J. Clin. Oncol.* **18**: 307–316.

106. Ringden, O., Uzunel, M., Rasmusson, I., Remberger, M., Sundberg, B., Lonnies, H., Marschall, H. U., Dlugosz, A., Szakos, A. et al. (2006). Mesenchymal stem cells for treatment of therapy-resistant graft-versus-host disease. *Transplantation* **81**: 1390–1397.

107. Koc, O. N., Gerson, S. L., Cooper, B. W., Dyhouse, S. M., Haynesworth, S. E., Caplan, A. I. & Lazarus, H. M. (2000). Rapid hematopoietic recovery after coinfusion of autologous-blood stem cells and culture-expanded marrow mesenchymal stem cells in advanced breast cancer patients receiving high-dose chemotherapy. *J. Clin. Oncol.* **18**: 307–316.

108. Parent, J. M. (2003). Injury-induced neurogenesis in the adult mammalian brain. *Neuroscientist* **9**: 261–272.

109. Fischbach, G. D. & McKhann, G. M. (2001). Cell therapy for Parkinson's disease. *N. Engl. J. Med.* **344**: 763–765.

110. Freed, C. R., Greene, P. E., Breeze, R. E., Tsai, W. Y., DuMouchel, W., Kao, R., Dillon, S., Winfield, H., Culver, S. et al. (2001). Transplantation of embryonic dopamine neurons for severe Parkinson's disease. *N. Engl. J. Med.* **344**: 710–719.

111. Gill, S. S., Patel, N. K., Hotton, G. R., O'Sullivan, K., McCarter, R., Bunnage, M., Brooks, D. J., Svendsen, C. N. & Heywood, P. (2003). Direct brain infusion of glial cell line-derived neurotrophic factor in Parkinson disease. *Nat. Med.* **9**: 589–595.

112. Eglitis, M. A. & Mezey, E. (1997). Hematopoietic cells differentiate into both microglia and macroglia in the brains of adult mice. *Proc. Natl. Acad. Sci. USA* **94**: 4080–4085.

113. Chopp, M. & Li, Y. (2002). Treatment of neural injury with marrow stromal cells. *Lancet Neurol.* **1**: 92–100.

114. Devine, S. M., Cobbs, C., Jennings, M., Bartholomew, A. & Hoffman, R. (2003). Mesenchymal stem cells distribute to a wide range of tissues following systemic infusion into nonhuman primates. *Blood* **101**: 2999–3001.

115. Cogle, C. R., Yachnis, A. T., Laywell, E. D., Zander, D. S., Wingard, J. R., Steindler, D. A. & Scott, E. W. (2004). Bone marrow transdifferentiation in brain after transplantation: a retrospective study. *Lancet* **363**: 1432–1437.

116. Weimann, J. M., Charlton, C. A., Brazelton, T. R., Hackman, R. C. & Blau, H. M. (2003). Contribution of transplanted bone marrow cells to Purkinje neurons in human adult brains. *Proc. Natl. Acad. Sci. USA* **100**: 2088–2093.

117. Gussoni, E., Bennett, R. R., Muskiewicz, K. R., Meyerrose, T., Nolta, J. A., Gilgoff, I., Stein, J., Chan, Y. M., Lidov, H. G. et al. (2002). Long-term persistence of donor nuclei in a Duchenne muscular dystrophy patient receiving bone marrow transplantation. *J. Clin. Invest.* **110**: 807–814.

118. Levy, Y. S., Stroomza, M., Melamed, E. & Offen, D. (2004). Embryonic and adult stem cells as a source for cell therapy in Parkinson's disease. *J. Mol. Neurosci.* **24**: 353–386.

119. Dass, B., Olanow, C. W. & Kordower, J. H. (2006). Gene transfer of trophic factors and stem cell grafting as treatments for Parkinson's disease. *Neurology.* **66**: S89–S103.

120. Li, Y., Chen, J., Wang, L., Zhang, L., Lu, M. & Chopp, M. (2001). Intracerebral transplantation of bone marrow stromal cells in a 1-methyl-4-phenyl-1,2,3,6-tetrahydropyridine mouse model of Parkinson's disease. *Neurosci. Lett.* **316**: 67–70.

121. Brederlau, A., Correia, A. S., Anisimov, S. V., Elmi, M., Paul, G., Roybon, L., Morizane, A., Bergquist, F., Riebe, I. et al. (2006). Transplantation of human embryonic stem cell-derived cells to a rat model of Parkinson's disease: effect of in vitro differentiation on graft survival and teratoma formation. *Stem Cells* **24**: 1433–1440.

122. Dunnett, S. B. & Rosser, A. E. (2007). Stem cell transplantation for Huntington's disease. *Exp. Neurol.* **203**: 279–292.

123. Rosser, A. E. & Dunnett, S. B. (2003). Neural transplantation in patients with Huntington's disease. *CNS Drugs* **17**: 853–867.

124. Philpott, L. M., Kopyov, O. V., Lee, A. J., Jacques, S., Duma, C. M., Caine, S., Yang, M. & Eagle, K. S. (1997). Neuropsychological functioning following fetal striatal transplantation in Huntington's chorea: three case presentations. *Cell Transplant.* **6**: 203–212.

125. Lescaudron, L., Unni, D. & Dunbar, G. L. (2003) Autologous adult bone marrow stem cell transplantation in an animal model of huntington's disease: behavioral and morphological outcomes. *Int. J. Neurosci.* **113**: 945–956.

126. Vazey, E. M., Chen, K., Hughes, S. M. & Connor, B. (2006). Transplanted adult neural progenitor cells survive, differentiate and reduce motor function impairment in a rodent model of Huntington's disease. *Exp. Neurol.* **199**: 384–396.

127. Prineas, J. W. & McDonald, W. I. (1997). Demyelinating diseases. In: Greenfield's

Neuropathology (Graham, D. I. & Lantos, P. L., eds.), pp. 813–895. Arnold, London.

128. Ferguson, B., Matyszak, M. K., Esiri, M. M. & Perry, V. H. (1997). Axonal damage in acute multiple sclerosis lesions. *Brain* **120**: 393–399.

129. Trapp, B. D., Peterson, J., Ransohoff, R. M., Rudick, R. A., Mork, S. & Bo, L. (1998). Axon transection in the lesions of multiple sclerosis. *N. Engl. J. Med.* **338**: 278–285.

130. Scolding, N. & Franklin, R. (1998). Axon loss in multiple sclerosis. *Lancet* **352**: 340–341.

131. Smith, K. J. & McDonald, W. I. (1999). The pathophysiology of multiple sclerosis: the mechanisms underlying the production of symptoms and the natural history of the disease. *Philos. Trans. R. Soc. Lond B Biol. Sci.* **354**: 1649–1673.

132. Lassmann, H., Bruck, W., Lucchinetti, C. F. & Rodriguez, M. (1997). Remyelination in multiple sclerosis. *Multiple Sclerosis* **3**: 133–136.

133. Prineas, J. W. & Connell, F. (1979). Remyelination in multiple sclerosis. *Ann. Neurol.* **5**: 22–31.

134. Raine, C. S. & Wu, E. (1993) Multiple sclerosis: remyelination in acute lesions. *J. Neuropathol. Exp. Neurol.* **52**: 199–204.

135. Patrikios, P., Stadelmann, C., Kutzelnigg, A., Rauschka, H., Schmidbauer, M., Laursen, H., Sorensen, P. S., Bruck, W., Lucchinetti, C. & Lassmann, H. (2006). Remyelination is extensive in a subset of multiple sclerosis patients. *Brain* **129**: 3165–3172.

136. Bruck, W., Kuhlmann, T. & Stadelmann, C. (2003). Remyelination in multiple sclerosis. *J Neurol Sci.* **206**: 181–185.

137. McDonald, W. I. (1996). Mechanisms of relapse and remission in multiple sclerosis. In: The Neurobiology of Disease (Bostock, H., Kirkwood, P. A. & Pullen, A. H., eds.), pp. 118–123. University Press, Cambridge.

138. Griffiths, I., Klugmann, M., Anderson, T., Yool, D., Thomson, C., Schwab, M. H., Schneider, A., Zimmermann, F., McCulloch, M. et al. (1998). Axonal swellings and degeneration in mice lacking the major proteolipid of myelin. *Science* **280**: 1610–1613.

139. Duncan, I. D., Paino, C., Archer, D. R. & Wood, P. M. (1992). Functional capacities of transplanted cell-sorted adult oligodendrocytes. *Dev. Neurosci.* **14**: 114–122.

140. Duncan, I. D., Grever, W. E. & Zhang, S. C. (1997). Repair of myelin disease: strategies and progress in animal models. *Mol. Med. Today* **3**: 554–561.

141. Franklin, R. J. M. & Blakemore, W. F. (1997). Transplanting oligodendrocyte progenitors into the adult CNS. *J. Anat.* **190**: 23–33.

142. Groves, A. K., Barnett, S. C., Franklin, R. J. M., Crang, A. J., Mayer, M., Blakemore, W. F. & Noble, M. (1993). Repair of demyelinated lesions by transplantation of purified O- 2A progenitor cells. *Nature* **362**: 453–455.

143. Rosenbluth, J. (1996). Glial transplantation in the treatment of myelin loss or deficiency. In: The Neurobiology of Disease: Contributions from Neuroscience to Clincial Neurology (Bostock, H., Kirkwood, P. A. & Pullen, A. H., eds.), pp. 124–148. Cambridge University Press, Cambridge, UK.

144. Warrington, A. E., Barbarese, E. & Pfeiffer, S. E. (1993). Differential myelinogenic capacity of specific developmental stages of the oligodendrocyte lineage upon transplantation into hypomyelinating hosts. *J. Neurosci. Res.* **34**: 1–13.

145. Kocsis, J. D. (1999). Restoration of function by glial cell transplantation into demyelinated spinal cord. *J. Neurotrauma* **16**: 695–703.

146. Baron-Van Evercooren, A., Avellana-Adalid, V., Lachapelle, F. & Liblau, R. (1997). Schwann cell transplantation and myelin repair of the CNS. *Mult. Scler.* **3**: 157–161.

147. Barnett, S. C., Franklin, R. J. M. & Blakemore, W. F. (1993). In vitro and in vivo analysis of a rat bipotential O-2A progenitor cell line containing the temperature-sensitive mutant gene of the SV40 large T antigen. *Eur. J. Neurosci.* **5**: 1247–1260.

148. Tontsch, U., Archer, D. R., DuboisDalcq, M. & Duncan, I. D. (1994). Transplantation of an oligodendrocyte cell line leading to extensive myelination. *Proc. Natl. Acad. Sci. USA* **91**: 11616–11620.

149. Utzschneider, D. A., Archer, D. R., Kocsis, J. D., Waxman, S. G. & Duncan, I. D. (1994). Transplantation of glial cells enhances action potential conduction of amyelinated spinal cord axons in the myelin- deficient rat. *Proc. Natl. Acad. Sci. USA* **91**: 53–57.

150. Jeffery, N. D., Crang, A. J., O'leary, M. T., Hodge, S. J. & Blakemore, W. F. (1999). Behavioural consequences of oligodendrocyte progenitor cell transplantation into experimental demyelinating lesions in the rat spinal cord. *Eur. J. Neurosci.* **11**: 1508–1514.

151. Zhang, J., Li, Y., Chen, J., Cui, Y., Lu, M., Elias, S. B., Mitchell, J. B., Hammill, L., Vanguri, P. & Chopp, M. (2005). Human bone marrow stromal cell treatment improves neurological functional recovery in EAE mice. *Exp. Neurol.* **195**: 16–26.

152. Akiyama, Y., Radtke, C., Honmou, O. & Kocsis, J. D. (2002). Remyelination of the spinal cord following intravenous delivery of bone marrow cells. *Glia* **39**: 229–236.

153. Akiyama, Y., Radtke, C. & Kocsis, J. D. (2002) Remyelination of the rat spinal cord by transplantation of identified bone marrow stromal cells. *J. Neurosci.* **22**: 6623–6630.

154. Sasaki, M., Honmou, O., Akiyama, Y., Uede, T., Hashi, K. & Kocsis, J. D. (2001). Transplantation of an acutely isolated bone marrow fraction repairs demyelinated adult rat spinal cord axons. *Glia* **35**: 26–34.

155. Scolding, N., Marks, D., Rice, C. (2007). Autologous mesenchymal bone marrow stem cells: Practical considerations. *J. Neurol. Sci.* **265**: 111–115.

156. Mazzini, L., Mareschi, K., Ferrero, I., Vassallo, E., Oliveri, G., Nasuelli, N., Oggioni, G. D., Testa, L. & Fagioli, F. (2008). Stem cell treatment in Amyotrophic Lateral Sclerosis. *J Neurol Sci.* **265**: 78.

157. Silani, V., Cova, L., Corbo, M., Ciammola, A. & Polli, E. (2004). Stem-cell therapy for amyotrophic lateral sclerosis. *Lancet* **364**: 200–202.

158. Guan, Y. J., Wang, X., Wang, H. Y., Kawagishi, K., Ryu, H., Huo, C. F., Shimony, E. M., Kristal, B. S., Kuhn, H. G. & Friedlander, R. M. (2007). Increased stem cell proliferation in the spinal cord of adult amyotrophic lateral sclerosis transgenic mice. *J. Neurochem.* **102**: 1125–1138.

159. Eglitis, M. A., Dawson, D., Park, K. W. & Mouradian, M. M. (1999). Targeting of marrow-derived astrocytes to the ischemic brain. *Neuroreport* **10**: 1289–1292.

160. Zhao, L. R., Duan, W. M., Reyes, M., Keene, C. D., Verfaillie, C. M. & Low, W. C. (2002). Human bone marrow stem cells exhibit neural phenotypes and ameliorate neurological deficits after grafting into the ischemic brain of rats. *Exp. Neurol.* **174**: 11–20.

161. Chen, J., Li, Y., Wang, L., Lu, M., Zhang, X. & Chopp, M. (2001). Therapeutic benefit of intracerebral transplantation of bone marrow stromal cells after cerebral ischemia in rats. *J. Neurol. Sci.* **189**: 49–57.

162. Chen, J., Li, Y., Wang, L., Zhang, Z., Lu, D., Lu, M. & Chopp, M. (2001). Therapeutic benefit of intravenous administration of bone marrow stromal cells after cerebral ischemia in rats. *Stroke* **32**: 1005–1011.

163. Zhang, Z. G., Zhang, L., Jiang, Q. & Chopp, M. (2002). Bone marrow-derived endothelial progenitor cells participate in cerebral neovascularization after focal cerebral ischemia in the adult mouse. *Circ. Res.* **90**: 284–288.

164. Shyu, W. C., Lin, S. Z., Chiang, M. F., Su, C. Y. & Li, H. (2006). Intracerebral peripheral blood stem cell (CD34+) implantation induces neuroplasticity by enhancing beta1 integrin-mediated angiogenesis in chronic stroke rats. *J Neurosci.* **26**: 3444–3453.

165. Hess, D. C., Hill, W. D., Martin-Studdard, A., Carroll, J., Brailer, J. & Carothers, J. (2002). Bone marrow as a source of endothelial cells and NeuN-expressing cells after stroke. *Stroke* **33**: 1362–1368.

166. Arvidsson, A., Collin, T., Kirik, D., Kokaia, Z. & Lindvall, O. (2002). Neuronal replacement from endogenous precursors in the adult brain after stroke. *Nat. Med.* **8**: 963–970.

167. Hicks, A. U., Hewlett, K., Windle, V., Chernenko, G., Ploughman, M., Jolkkonen, J., Weiss, S. & Corbett, D. (2007). Enriched environment enhances transplanted subventricular zone stem cell migration and functional recovery after stroke. *Neuroscience* **146**: 31–40.

168. Li, Y., Chen, J., Chen, X. G., Wang, L., Gautam, S. C., Xu, Y. X., Katakowski, M., Zhang, L. J., Lu, M. et al. (2002). Human marrow stromal cell therapy for stroke in rat: neurotrophins and functional recovery. *Neurology* **59**: 514–523.

169. Kondziolka, D., Wechsler, L., Goldstein, S., Meltzer, C., Thulborn, K. R., Gebel, J., Jannetta, P., DeCesare, S., Elder, E. M. et al. (2000). Transplantation of cultured human neuronal cells for patients with stroke. *Neurology* **55**: 565–569.

170. Bang, O. Y., Lee, J. S., Lee, P. H. & Lee, G. (2005). Autologous mesenchymal stem cell transplantation in stroke patients. *Ann. Neurol.* **57**: 874–882.

171. Sprigg, N., Bath, P. M., Zhao, L., Willmot, M. R., Gray, L. J., Walker, M. F., Dennis, M. S. & Russell, N. (2006). Granulocyte-colony-stimulating factor mobilizes bone marrow stem cells in patients with subacute ischemic stroke: the stem cell trial of recovery enhancement after stroke (STEMS) pilot randomized, controlled trial (ISRCTN 16784092). *Stroke* **37**: 2979–2983.

Chapter 3

Human Trials for Neurodegenerative Disease

Claire M. Kelly, O.J. Handley, and A.E. Rosser

Summary

The lack of disease-modifying treatments currently available for not just some but most neurodegenerative diseases, including Parkinson's disease, Huntington's disease, and even stroke, helps explain increasing interest in cell-based therapies. One key aim of such treatment is to replace neurons or glia lost as a result of the disease, with a view to the cells integrating functionally within the host tissue in order to reconstruct neural circuitry. Clinical trials using primary human fetal tissue as a cell source commenced in Parkinson's disease (PD) in the 1980s; currently, comparable neural transplantation trials in Huntington's disease are underway. Disappointing results of later controlled trials in PD illustrated not least the vital importance of methodological issues relating to the structure and implementation of clinical trials, and these issues will be considered here in more depth.

Key words: Neural transplantation, Huntington's disease, Parkinson's disease, Stem cell therapy

1. Introduction

Human neurodegenerative disorders such as Parkinson's disease (PD), Huntington's disease (HD), multiple sclerosis (MS), and stroke are caused by a loss of neurons and/or glia. Treatment of these devastating neurological diseases is limited and the lack of a clear mechanism by which such diseases are manifest restricts the treatment modalities that can be considered, as specific targets are not yet identified. Thus, disease-modifying treatments are not yet available for most of these conditions. In the light of this there is considerable interest in neural transplantation as a potential therapeutic strategy. The aim of neural transplantation is to replace the neurons or glia that have been lost as a result of the disease with a view to the cells integrating and synapsing, in the case of

Neil J. Scolding and David Gordon (eds.), *Methods in Molecular Biology, Neural Cell Transplantation vol. 549*
DOI: 10.1007/978-1-60327-931-4_3, © Humana Press, a part of Springer Science + Business Media, LLC 2009

neurons, with the host tissue in order to reconstruct the remaining host circuitry. In addition to this, implanted cells may also be used to supply factors such as growth factors or for gene therapy. In the case of stroke models, implantation of certain stem cell types appears to function by encouraging mobilisation of the endogenous stem cell pool.

To date, the most success in terms of achieving circuit reconstruction has been when donor cells are procured from primary foetal tissue obtained from elective termination of pregnancy. However, the logistical and ethical issues associated with the use of this tissue have prevented widespread application. Therefore, alternative donor cell sources are being actively explored, one such cell source being stem cells. Stem cells have the potential to self-propagate, thus generating large numbers of cells, whilst retaining the capacity to differentiate into mature phenotypes. These properties make them an attractive renewable potential source of donor cells.

2. Stem and Precursor Cells as Potential Donor Cell Sources (Fig. 1)

There are many different definitions and categories of stem cell, leading to a somewhat confusing nomenclature. An embryonic stem (ES) cell is derived from the inner cell mass of the blastocyst and has the potential to differentiate into any cell type of the body. Significant ethical disputes have been associated with the derivation and use of ES cells, including concerns over the use of human embryos and fears related to human cloning. As a result of these ethical issues, many countries have restricted or banned ES cell research. Nevertheless, other countries have actively supported the development of ES cell research because of the perceived therapeutic benefit in various diseases. Some countries, such as the United Kingdom, allow cloning of human embryos for therapeutic purposes but impose tight regulations to preclude their use for reproductive cloning.

ES cells can be expanded in culture to generate large populations of cells. Differentiation to tissue-specific lineages can be achieved by exogenous means. For example, ES cell proliferation is maintained in the presence of serum and LIF (leukaemia inhibitory factor), but on removal of serum and LIF from the media the cells have been found to default to a neural lineage. The precise mechanisms that guide the differentiation of a stem cell towards a specific mature phenotype depend on the phenotype in question. For example, for dopaminergic neurons to be generated it has been shown that sonic hedgehog (Shh) and retinoic

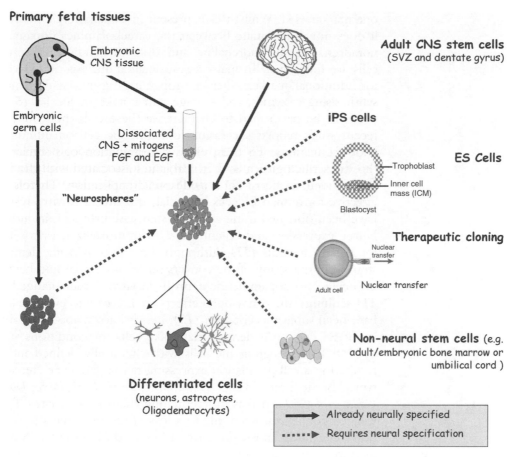

Fig. 1. A schematic that illustrates the major sources of stem cells under consideration for production of donor cells for neural transplantation. Stem cells derived from the developing brain are already lineage restricted to produce cells of a neural phenotype, but most other sources need to be manipulated to produce neural cells, and other cell types need to be eliminated. For all stem cell sources (including neural stem cells after more than 1–2 passages), directing the cells to the precise neural phenotype needed for use in specific neurodegenerative conditions remains a major challenge.

acid (RA) are amongst a number of important factors, but that the doses and timing of exposure of the cells to these factors are important. Learning to recapitulate in vitro such events in order to "direct" the differentiation of these cells to a particular cell type is complex.

However, there has been progress in achieving specific dopaminergic neuron differentiation of ES-derived neurons with reports of 16–35% tyrosine hydroxylase (TH) positive neurons being generated by the addition of specific factors to the culture medium (1–4). Expression of the transcription factor Nurr1 enhances the differentiation of ES cells into dopaminergic neurons with reports of 80% TH-positive neurons being generated (5–9). However, many of these studies have based their results on the expression of one marker, TH. Whilst this is present in dopaminergic neurons it does not differentiate between the catecholamines dopamine, noradrenaline, and adrenaline and does not indicate that the cells are functional. In many cases, detailed analysis of TH cells for additional markers such as dopamine receptors, dopamine synthesising enzymes, and storage and uptake molecules *(5, 7)* needs to be performed to characterise these cells more fully. A recent study reports successful dopaminergic neuron differentiation from mouse ES cells, using step-wise protocols similar to those described earlier, both in vitro and following transplantation into a rodent model of PD *(10)*. Proving that similar protocols are effective for human ES cells is crucial, and similar in vitro results have been obtained using a similar step-wise protocol, including Nurr1 expression and co-culture with immortalized asytocytes, for human ES cells *(11)*. Although there is convincing demonstration in this study of the conversion of substantial numbers of hES cells into dopaminergic neurons in vitro (largely gauged by TH staining), the behavioural effects are less easy to evaluate and have been subject to criticism *(12)*. Less is known about the ability of ES cells to generate striatal-like cells for conditions such as HD. Differentiating the cells with chemically defined media resulted in a cell population expressing neural fate characteristics typical of the forebrain such *Dlx5, Dlx1, Lhx5, Tbr1, Pax6, Dbx1, Gsh2, and Gsh1* whereas differentiating them to alternative fates was temporally restricted due to a loss of responsiveness to positional cues *(13)*. In the presence of FGF-2 during the first 8 days in culture these cells maintain a largely neuronal fate. However, with successive passaging an ontogenic drift towards gliogenesis is evident *(14)*. Another important issue is the potential of ES cells to form tetracarcinomas, since remaining undifferentiated ES cells in grafted cell suspension can continue to divide forming tumours. For example, Bjorklund et al. grafted a mouse ES cell line into a rat model of PD and reported that 5 out of 25 grafts formed teratoma-like tumours with resulting death of the animals *(15)*. One method for eliminating undifferentiated cells is by the introduction of suicide genes, such as the *E. coli gpt* and herpes thymidine kinase (*HSVtk*), into the cells prior to transplantation. The plasmid vector also contained a neomycin resistance gene that allowed selection of the undifferentiated ES cells, as differentiated cells in the presence of the neomycin resistance gene would be resistant to the effects of ganciclovir. Undifferentiated HSVtk-positive cells that continue to proliferate can then be destroyed by the conversion of prodrug nucleoside ganciclovir to its phosphorylated form, which is then incorporated into the DNA of replicating cells resulting in apoptosis of the cells *(16, 17)*. The functionality and efficacy of the differentiated cells will also have to be addressed as well as the possibility of rejection before they can be considered for clinical trials.

Induced pluripotent stem (iPS) cells are a pluripotent population of cells that can be derived from adult somatic cells. They were first produced in mouse in 2006 *(18)* and subsequently in human in 2007 *(19, 20)* by transducing adult and embryonic fibroblasts with defined factors such as Nanog, Oct4 and Sox2. These cells have presented with characteristics that are very similar if not identical to ES cells and they are an attractive alternative cell source for cell based therapies given that they can be derived from the patients own fibroblast, thereby overcoming the ethical issues associated with the use of ES cells. Much work is required to identify ways of isolating these cells without the need for modification and to further develop these cells for clinical use (for review see *(21)*).

Tissue-specific stem cells such as neural stem cells are derived from neural tissue and are therefore already committed to a neural lineage. It is not clear in all cases that such cells retain the properties of a true stem cell when cultured in vitro and it may be considered more correct to refer to these cells as neural precursor cells. The restriction of differentiated neural stem cells to phenotypes of a neural lineage makes them interesting as a donor cell source for CNS transplantation. Neural precursor cells expand in vitro in the presence of growth factors such as FGF-2 and EGF and form free-floating spheres of cells referred to as neurospheres. These cells have the potential to differentiate into the three cell types of the central nervous system: neurons, glia, and oligodendrocytes. However, such cells have proven less easy to direct towards a specific phenotype than have ES cells *(22–24)* and a further problem with neural stem cells is that they appear to have limited proliferative and neurogenic potential, which may restrict their therapeutic application *(25, 26)*.

3. Clinical Trials

Clinical trials of primary human foetal tissue (i.e., developing cells taken directly from the foetal brain without any manipulation or expansion in culture) are important in paving the way for trials of stem and precursor cell transplantation by demonstrating that cell replacement can be safe and effective in neurodegenerative disease. Foetal cells from the appropriate brain region (e.g. the ventral mesencephalic region for PD) are already committed to the required phenotype and so can provide proof of principle that the cell replacement approach is valid. Clinical trials are ongoing for PD and early trials are now underway for HD based on extensive experimental work in animal models. The PD trials that began in the late 1980s used primary human foetal mesencephalic tissue as the host tissue and transplanted it into the host striatum, which is the target area of these cells. Placing a graft into the substantia nigra is not viable, as the cells are unable to reliably project the distance from the substantia nigra to their targets in the striatum. The mesencephalic tissue contains fate-committed dopaminergic neuroblasts, which have the capacity to differentiate into fully mature dopaminergic neurones following transplantation, provided that the biological principles arising from animal work are adhered to. These include harvesting tissue between specific gestational ages and the optimisation of tissue preparation methodologies as described earlier. If one considers the PD trials in which these principles are taken into account and which use good longitudinal assessment, then results to date in the PD trials have demonstrated improvements in a range of motor skills in some patients, and many, but not all, of the patients have been able to reduce or even eliminate their daily intake of l-dopa *(27, 28)*. However, there is variability in the success of this approach, which may be a direct result of variations in transplant methodology as well as differences in patient selection criteria *(29–31)*. Recent trials have also highlighted the possibility of dyskinetic side effects in some patients *(29)*, and the reason for these is currently a topic of active investigation *(32)*.

Parallel clinical trials of neural transplantation in HD are at a much earlier stage than the PD trials and are currently underway in a small number of centres around the world. The French trial, based in Créteil, was the first to provide efficacy data, based on systematic long-term evaluation of their patients *(33)*. Three of the five patients, having received bilateral striatal implants, were reported to show substantial improvement over several years *(34)*. More recently there has been an expansion of the French trial to include other French-speaking regions in Europe and a total of 40 patients have now received transplants and are currently undergoing follow-up with no efficacy data available as yet. In another study in Florida, 6 of 7 patients appeared to show improvement but one declined significantly, so that the overall group changes were not significant *(35)*. One patient died after 18 months due to cardiovascular disease, and postmortem analysis of this patient's brain showed surviving graft tissue with evidence of striatal cell differentiation and appeared to be unaffected by the underlying disease progression and free of immune attack *(30, 36)*. In the same study three patients developed subdural haemor-rhages and two required surgical drainage probably due to the relatively advanced disease stage in this study compared with others *(35)*. Small numbers of patients have received grafts in several other centres with reports of safety *(37, 38)*, and although efficacy studies are underway in these centres, systematic reports have not yet been published.

The initial studies of cell transplantation in PD and HD are providing accumulating evidence of the conditions for safety and preliminary evidence for efficacy. However, the limited availability of foetal tissue and the difficulty in ensuring the high degree of standardisation and quality control when a continuous source of fresh donor tissue is required from elective abortion limits the widespread use of neural transplantation as a practical therapy. Ethical and legislative concerns about abortion and the large number of donors required to support each operation already restrict the number of patients who can receive grafts to a few specialist centres in a restricted number of countries. Moreover, the shifting preference for medical rather than surgical abortions may further limit the availability of tissues to supply even the limited number of programmes already in progress. These issues have stimulated the search for alternative sources of donor cells or tissue that circumvent the problems associated with primary foetal tissue collection.

To date there have been no reported systematic clinical studies of stem/precursor cell transplants in PD or HD, although there have been clinical studies of a variety of renewable cell sources in other neurodegenerative diseases. For example, phase I and phase II clinical trials for stroke have commenced using hNT cells and immunosuppressive treatment for 8 weeks. These cells are post-mitotic immature neurons generated from the immortalized NT2 cell line derived from a human teratocarcinoma *(39)*. The trial was a very small one with only 14 patients receiving transplants in the phase II trial and therefore efficacy cannot be taken from the data. However, it was noted that only 6 patients of the 14 receiving transplants showed improvement on the European stroke scale albeit a non-significant improvement, although significant improvements were seen on some of the secondary neurological outcome measures *(40–43)*. From this study further trials can be initiated taking into account the lessons learnt. Bone marrow-derived stromal cells have also been used in a clinical trial; however, in this instance the cells were infused intravenously and a slight improvement on the patients clinical scale was reported *(44)*. More recently HUCB cells have been approved for clinical trials in traumatic brain injury in children. Xenotransplantation of porcine-derived neural precursor cells into patients with stroke has also been carried out. In this instance 4 years after transplantation 2 of the 5 patients showed functional improvement but only one was significant. One major difference between this and the previous hNT trial is that no immunosuppression was administered; instead, the cells were pre-treated with an anti-major histocompatibility complex class 1 antibody based on preclinical animal experiments.

4. Preparation of Cells for Clinical Trials

Prior to the commencement of clinical trials it is imperative that there are sufficient pre-clinical data to back up the progression to clinical trials. In the case of neural transplantation the pre-clinical data need to be of the highest quality and verified in more than one lab. We consider transplantation experiments in animal models of the disease in question, to demonstrate graft survival, appropriate graft differentiation, and convincing evidence of behavioural improvement to be essential before considering clinical application. Having accumulated sufficient data, all steps in the process of cell preparation are validated under good manufacturing practice (GMP) conditions. This is a prerequisite of all cell preparations for human use under the EU directive; see later. Therefore, all processes in the preparation of a tissue sample for transplantation must be carried out using specified products that have been tested and validated for the purpose. The process itself must be carried out in a GMP facility, which is equivalent to a pharmaceutical grade clean room and is therefore governed by strict rules and regulations all of which are outlined in the directive and which are licensed in the UK by the HTA (human tissue authority). Effectively, this ensures that the tissue/cell source being used for transplantation is handled in the cleanest possible way, thereby reducing as far as possible the risk of any contamination to the cells prior to the transplantation procedure. Regular checks are carried out to ensure that there has been no change to the degree of cleanliness and where necessary steps are taken to rectify any issues. This involves regular decontamination of the facility and measuring air pressure and particle levels within and around the working area all of which must be clearly documented.

Having established a cell source that can be maintained under GMP conditions, the cells may then be deposited to the stem cell bank for long-term storage (http://www.ukstemcellbank.org.uk/). The UK stem cell bank is funded by the MRC and is not involved in stem cell biology research to avoid conflicts of interest. Cell lines deposited to the bank are well characterised, and further characterisation where necessary can be carried out by the stem cell bank itself.

Clinical trials of cell therapy must follow frameworks set out by clinical and research governance to ensure the protection and promotion of the health and well-being of the subject at all times. Successful frameworks embody clear lines of responsibility and accountability, the implementation of the highest possible standards of clinical care and a constant dynamic of improvement. There are a large number of regulatory organisations and systems in place to authorise, manage, and review the conduct of clinical trials across countries. The following section provides an overview of current regulations, guidelines and authorities employed to ensure the proper conduct of clinical trials of cell therapy.

A European Union (EU) directive is a legislative act that applies to EU member states (currently 27 European countries). A directive requires the member states to achieve a particular result without laying down the means of achieving that result. Here, we describe two EU directives that are relevant to clinical trials of cell therapy.

5. EU Directive 2001/20/EC Relating to the Implementation of Good Clinical Practice in the Conduct of Clinical Trials on Medicinal Products for Human Use

The EU directive 2001/20/EC came into force on May 1, 2004. The purpose of this directive is to enhance the protection of human subjects enrolled in clinical trials of medicinal products, to ensure quality of conduct, and to standardise regulation and conduct of clinical trials across Europe. Paramount to the purpose of a clinical trial is to establish the safety and efficacy of a therapeutic intervention using one or more medicinal products. All trials must be designed, conducted, and reported in accordance with the principles of good clinical practice (GCP), ensuring the protection of human rights and dignity. The directive sets out the laws, regulations, and administrative procedures that must be adhered to when undertaking clinical trials of medicinal products for human use. Each clinical trial must obtain a single favourable opinion from a legally constituted Research Ethics Committee. During the conduct of a clinical trial, statutory inspections should be conducted by authorities able to verify acceptable levels of efficacy and safety of medicines and medical devices, and compliance with the standards of GCP. In the UK a government agency known as the Medicines and Healthcare products Regulatory Agency (MHRA) administers, issues and reports on the conduct of clinical trials. Clinical trials must follow regulated procedures to monitor adverse reactions, known as pharmacovigilance. This system activates the immediate cessation of a trial with an unacceptable level of risk and exchanges information between Member States using Eudravigilance and EudraCT (a database of al clinical trials commencing in the EU from May 1, 2004). Each clinical trial of an investigational medicinal product (IMP) is required to obtain a EudrACT number. Given the potential adverse reactions to any IMP, provisions should be made for insurance or indemnity to cover the liability of the investigator and sponsor.

6. EU DIRECTIVE 2004/23/EC on Setting Standards of Quality and Safety for the Donation, Procurement, Testing, Processing, Preservation, Storage, and Distribution of Human Tissues and Cells

The use of human tissues and cells for transplantation has become a strong focus for therapeutic intervention within neurodegenerative diseases. The Directive only applies to the use of human tissue and cells used for clinical trials to the human body; it does not apply to in vitro research experiments or use in animal models of disease. It is essential that the quality and safety of these substances is maintained, in particular to prevent the transmission of diseases. Donated tissues and cells must be procured, tested, processed, preserved and stored in accordance with validated and approved safety measures. A unified framework of standard operating procedures (SOPs) is required to regulate the highest quality of tissue and cells for use in transplantation. The procurement and testing of tissue and cells should be executed by appropriately trained and experienced individuals within laboratory conditions that have received accreditation, authorisation, and licensing by the regulatory body. To promote the highest possible standard of tissue and cells for use in human clinical trials, the Directive insists that the handling, storage, and preparation of substances meets with GMP. GMP refers to highly specified conditions of laboratory practice and regulation put in place to ensure successful production and quality control of the therapeutic product.

In light of the EU directive, the UK Human Tissue Act (2004) has set out a new legal framework for the storage and use of tissue from the living, and the removal, storage, and use of tissues and organs from the dead. The Human Tissue Act also covers residual tissue that has been obtained following clinical and diagnostic procedures. In the UK, the HTA is the regulatory body for all matters concerning the removal, storage, use, and disposal of human tissue for scheduled purposes. The HTA (www.hta.gov.uk) provides advice and guidance on the Human Tissue Act, issues codes of practice and is responsible for licensing establishments. The fundamental principle of the HTA is that of consent in relation to the retention and use of living patients' organs and tissue for particular purposes beyond their diagnosis and treatment, and consent surrounding the removal, retention, and use of tissue from those who have died (where consent is obtained either by those individuals in life, or after death by someone nominated by or close to them). The codes of practice are guidelines that should be applied and adhered to by individuals undertaking relevant activities. The only exception where an HTA licence is not required is where tissue is being stored for use in a specific ethically approved project and is not retained after that project for unspecified future use. The licence permits the specified activity to take place within a specified location under the supervision of the designated individuals (DI) or corporate body named on the licence. The DI is responsible for ensuring that delegated individuals adhere to regulated SOPs. The HTA commands that host organizations are responsible for obtaining consent, ethical approval, ensuring anonymity where necessary, review whether a license is applicable, consolidate holdings, liaise with licensed tissue banks, and handle surplus tissue appropriately.

7. The Declaration of Helsinki and GCP

The Declaration of Helsinki was introduced by the World Medical Association 'as a statement of ethical standards to provide guidance to physicians and other participants in medical research involving human subjects'. It is the responsibility of the physician to promote and protect the health and rights of the human subject. The impetus of medical research on human subjects should aim to contribute to one of the following areas of investigation: enhance diagnosis, management, treatment or prevention of the disease, to understand the aetiology and the pathogenesis of the disease. Study investigators are expected to be conversant with ethical, legal, and regulatory requirements for research on human subjects in their own countries as well as relevant international requirements.

GCP is an internationally harmonised standard based on the Declaration of Helsinki. It is a standard of practice that should be integrated into clinical investigations involving the participation of human subjects. The guidelines set out the responsibilities, procedures, and processes that should be integrated into clinical trials to assure ethical and scientific integrity, whilst also serving to protect the rights, safety, and well-being of the subjects. Finally, compliance with GCP recommendations maximises the credibility and accuracy of the data extrapolated from the trial.

8. Independent Review Board

According to the Declaration of Helsinki research that includes experimental procedures involving human subjects should be 'submitted for consideration, comment, guidance, and where appropriate, approval to a specially appointed ethical review committee, which must be independent of the investigator, the sponsor or any other kind of undue influence'. GCP recommends that the independent review board (IRB) should be an independent body (a review board or a committee, institutional, regional, national or supranational), consisting of medical professionals and non-medical members who collectively have the qualifications and experience to assess the scientific, medical, and ethical aspects of the proposed trial. The responsibility of the IRB is to protect the rights, safety, and well-being of human subjects involved in a trial. The committee is expected to review and approve the trial protocol following consideration of, among other things, the suitability of the investigator(s), site facilities, and the procedures to collect and retain data, and also the processes in place to obtain and document informed consent of the trial subjects. IRBs should conform to the laws and regulations of the country in which the research is performed.

9. Outcome Measures

Cell therapy in neurodegenerative disease aims to restore brain circuitry and function by surgical replacement of the cells lost as a consequence of progressive neurodegenerative disease. The clinically beneficial effect of cell therapy is dependent on two main issues: first, whether the neural graft will survive and replace lost neurons in the place(s) where required, and second, whether the neural graft can integrate into the host neuronal circuitry and restore normal function. Assessing the true efficacy of any cell therapy depends on sensitive and reliable outcome measures. Ideally, an assessment battery to assess efficacy of stem cell therapy would utilise clinical and biological markers collected systematically over a longitudinal period. To date, previous and on-going clinical trials of primary tissue cell therapy in HD and PD have relied on the CAPIT-HD and CAPIT-PD protocols, respectively (CAPIT: Core Assessment Protocol for Intracerebral Transplantation), capturing longitudinal data on clinical and imaging markers. In spite of these existing protocols, there remains a continued demand for the development of more accurate and reliable outcome measures to assess the scientific and economic feasibility of future cell-based therapy.

A key aspect to the successful execution and completion of a clinical trial relies on measures of quality control and quality assurance. Quality control refers to measures and procedures that are put in place in order to ensure that accuracy and reliability of data, whereas quality assurance refers to the evaluation of the effectiveness of these procedures. Quality control can be implemented through training and administration of SOPs, and quality assurance is maintained through monitoring of data to ensure adherence to GCP (see later), to determine the quality and plausibility of data, and compliance to the study protocol.

The purpose of monitoring clinical trials is twofold: (a) to assure that protection, safety and rights of the subjects are maintained, and (b) to assure the quality and integrity of the data. A trial sponsor carries the responsibility of monitoring and often delegates the task to a contract research organisation. Monitors are required to verify that the study investigators are conversant with the study protocol and plan, competent to execute a well-controlled study with adequate facilities, population sample and time, and to ensure that investigators understand their obligation to obtain the relevant IRB approvals, and informed consent. A key area of importance surrounds monitoring and reporting of adverse events. All clinical trials carry risk to the subject's safety, and the procedures in place to handle an adverse event or reaction are determined by its severity. In cases of serious events or reactions, investigators must execute a series of actions with strict adherence to SOPs in order to alert and register the event with the appropriate healthcare authorities.

10. Useful Sources of Information

Medicines and Healthcare products Regulatory Agency: www.mhra.gov.uk

Eudract: http://eudract.emea.europa.eu

Clinical Trials Toolkit: www.ct-toolkit.ac.uk

European Medicines Agency: www.emea.europa.eu

EU Clinical Trials Directive: www.eortc.be/Services/Doc/clinical-EU-directive-04-April-01.pdf

EU Tissue Directive: http://europa.eu.int/eurlex/pri/en/oj/dat/2004/l_102/l_1020040407en00480058.pdf

References

1. Kawasaki H, Mizuseki K, Nishikawa S, Kaneko S, Kuwana Y, Nakanishi S, Sasai Y. (2000). Induction of midbrain dopaminergic neurons from ES cells by stromal cell-derived inducing activity. *Neuron* 28:31–40.

2. Lee S-H, Lumelsky N, Studer L, Auerbach JM, McKay R. (2000). Efficient generation of midbrain and hindbrain neurons from mouse embryonic stem cells. *Nature Biotechnology* 18:675–9.

3. Okabe S, Forsberg-Nilsson K, Spiro AC, Segal M, McKay RDG. (1996). Development of neuronal precursor cells and functional postmitotic neurons from embryonic stem cells in vitro. *Mechanisms of Development* 59:89–102.

4. Rolletschek A, Chang H, Guan KM, Czyz J, Meyer M, Wobus AM. (2001). Differentia-tion of embryonic stem cell-derived dopaminergic neurons is enhanced by survival-promoting factors. *Mechanisms of Development* 105:93–104.

5. Chung S, Sonntag KC, Andersson T, Bjorklund LM, Park JJ, Kim DW, Kang UJ, Isacson O, Kim KS. (2002). Genetic engineering of mouse embryonic stem cells by Nurr1 enhances differentiation and maturation into dopaminergic neurons. *European Journal of Neuroscience* 16:1829–38.

6. Grothe C, Timmer M, Scholz T, Winkler C, Nikkhah G, Claus P, Itoh N, Arenas E. (2004). Fibroblast growth factor-20 promotes the differentiation of Nurr1-overexpressing neural stem cells into tyrosine hydroxylase-positive neurons. *Neurobiology of Disease* 17:163–70.

7. Kim JH, Auerbach JM, Rodriguez-Gomez JA, Velasco I, Gavin D, Lumelsky N, Lee SH, Nguyen J, Sanchez-Pernaute R, et al. (2002). Dopamine neurons derived from embryonic stem cells function in an animal model of Parkinson's disease. *Nature* **418**:50–6.

8. Kim JY, Koh HC, Lee JY, Chang MY, Kim YC, Chung HY, Son H, Lee YS, Studer L, et al. (2003). Dopaminergic neuronal differentiation from rat embryonic neural precursors by Nurr1 overexpression. *Journal of Neurochemistry* **85**:1443–54.

9. Wagner J, Akerud P, Castro DS, Holm PC, Canals JM, Snyder EY, Perlmann T, Arenas E. (1999). Induction of a midbrain dopaminergic phenotype in Nurr1-overexpressing neural stem cells by type 1 astrocytes. *Nature Biotechnology* **17**:653–9.

10. Rodriguez-Gomez JA, Lu JQ, Velasco I, Rivera S, Zoghbi SS, Liow JS, Musachio JL, Chin FT, Toyama H, et al. (2007). Persistent dopamine functions of neurons derived from embryonic stem cells in a rodent model of Parkinson disease. *Stem Cells* **25**:918–28.

11. Roy NS, Cleren C, Singh SK, Yang L, Beal MF, Goldman SA. (2006). Functional engraftment of human ES cell-derived dopaminergic neurons enriched by coculture with telomerase-immortalized midbrain astrocytes. *Nature Medicine* **12**:1259–68.

12. Christophersen NS, Brundin P. (2007). Large stem cell grafts could lead to erroneous interpretations of behavioral results?. *Nature Medicine* **13**:118; author reply 119.

13. Bouhon IA, Kato H, Chandran S, Allen ND. (2005). Neural differentiation of mouse embryonic stem cells in chemically defined medium. *Brain Research Bulletin 15* **68**:62–75.

14. Bouhon IA, Joannides A, Kato H, Chandran S, Allen ND. (2006). Embryonic stem cell-derived neural progenitors display temporal restriction to neural patterning. *Stem Cells* **24**:1908–13.

15. Bjorklund LM, SÂnchez-Pernaute R, Chung S, Andersson T, Ching Chen IY, McNaught KSP, Brownell AL, Jenkins BG, Wahlested C, et al. (2002). Embryonic stem cells develop into functional dopaminergic neurons after transplantation in a Parkinson rat model. *Proceedings of National Academy of Science USA* **99**:2344–9.

16. Fareed MU, Moolten FL. (2002). Suicide gene transduction sensitizes murine embryonic and human mesenchymal stem cells to ablation on demand – A fail-safe protection against cellular misbehavior. *Gene Therapy* **9**:955–62.

17. Schuldiner M, Itskovitz-Eldor J, Benvenisty N. (2003). Selective ablation of human embryonic stem cells expressing a "suicide" gene. *Stem Cells* **21**:257–65.

18. Thakahashi K, Yamanaka S. (2006). Induction of pluripotent stem cells from mouse embryonic and adult fibroblast cultures by defined factors. *Cell* **126**:663–676.

19. Takahashi K, Tanabe K, Ohnuki M, Narita M, Ichisaka T, Tomoda K, Yamanaka S. (2007). Induction of pluripotent stem cells from adult human fibroblasts by defined factors. *Cell* **131**:861–872.

20. Yu J, Vodyanik MA, Smuga-Otto K, Antosiewicz-Bourget J, Frane JL, Tian S, Nie J, Jonsdottir GA, Ruotti V, Stewart R, Slukvin, II, Thomsen JA. (2007). Induced pluripotent stem cell lines derived from human somatic cells. *Science* **318**:1917–1920.

21. Zhao R, Daley GQ. (2008). From fibroblasts to iPS cells: induced pluripotency by defined factors. *J Cell Biochem* **105**:949–955.

22. Cesnulevicius K, Timmer M, Wesemann M, Thomas T, Barkhausen T, Grothe C. (2006). Nucleofection is the most efficient nonviral transfection method for neuronal stem cells derived from ventral mesencephali with no changes in cell composition or dopaminergic fate. *Stem Cells* **24**:2776–91.

23. Anderson L, Burnstein RM, He X, Luce R, Furlong R, Foltynie T, Sykacek P, Menon DK, Caldwell MA. (2007). Gene expression changes in long term expanded human neural progenitor cells passaged by chopping lead to loss of neurogenic potential in vivo. *Experiments in Neurology* **204**:512–24.

24. Kim HJ, Sugimori M, Nakafuku M, Svendsen CN. (2007). Control of neurogenesis and tyrosine hydroxylase expression in neural progenitor cells through bHLH proteins and Nurr1. *Experiments in Neurology* **203**:394–405.

25. Andersson E, Jensen JB, Parmar M, Guillemot F, Bjorklund A. (2006). Development of the mesencephalic dopaminergic neuron system is compromised in the absence of neurogenin 2. *Development* **133**:507–16.

26. Zietlow R, Pekarik V, Armstrong RJ, Tyers P, Dunnett SB, Rosser AE. (2005). The survival of neural precursor cell grafts is influenced by in vitro expansion. *Journal of Anatomy* **207**:227–40.

27. Hagell P, Brundin P. (2001). Cell survival and clinical outcome following intrastriatal transplantation in Parkinson disease. *Journal of Neuropathology and Experimental Neurology* **60**:741–52.

28. Olanow CW, Kordower JH, Freeman TB. (1996). Fetal nigral transplantation as a therapy for Parkinson's disease. *Trends in Neurosciences* **19**:102–9.

29. Freed CR, Greene PE, Breeze RE, Tsai WY, DuMouchiel W, Kao R, Dillon S, Winfield H, Culver S, et al. (2001). Transplantation of

embryonic dopamine neurons for severe Parkinson's disease. *The New England Journal of Medicine* **344**:710–19.

30. Freeman TB, Cicchetti F, Hauser RA, Deacon TW, Li XJ, Hersch SM, Nauert GM, Sanberg PR, Kordower JH, et al. (2000). Transplanted fetal striatum in Huntington's disease: Phenotypic development and lack of pathology. *Proceedings of the National Academy of Sciences* **97**:13877–82.

31. Lindvall O, Brundin P, Widner H, Rehncrona S, Gustavii B, Frackowiak R, Leenders KL, Sawle G, Rothwell JC, et al. (1990). Grafts of fetal dopamine neurons survive and improve motor function in Parkinson's disease. *Science* **247**:574–7.

32. Hagell P, Piccine P, Björklund A, Brundin P, Rehncrona S, Widner H, Crabb L, Pavese N, Oertel WH, et al. (2002). Dyskinesias following neural transplantation in Parkinson's disease. *Nature Neuroscience* **5**:627–8.

33. Bachoud-Levi AC, Gaura V, Brugieres P, Lefaucheur JP, Boisse MF, Maison P, Baudic S, Ribeiro MJ, Bourdet C, et al. (2006). Effect of fetal neural transplants in patients with Huntington's disease 6 years after surgery: A long-term follow-up study. *Lancet Neurology* **5**:303–9.

34. Bachoud-Levi AC, Hantraye P, Peschanski M. (2002). Fetal neural grafts for Huntington's disease: A prospective view. *Movement Disorders* **17**:439–44.

35. Hauser RA, Sandberg PR, Freeman TB, Stoessl AJ. (2002). Bilateral human fetal striatal transplantation in Huntington's disease. *Neurology* **58**:687–685.

36. Freeman TB, Hauser RA, Sanberg PR, Saporta S. (2000). Neural transplantation for the treatment of Huntington's disease. *Progress in Brain Research* **127**:405–11.

37. Kopyov OV, Jacques S, Eagle KS. (1998). Fetal transplantation for the treatment of neurodegenerative diseases – Current status and future potential. *CNS Drugs* **9**:77–83.

38. Rosser AE, Barker RA, Harrower T, Watts C, Farrington M, Ho AK, Burnstein RM, Menon DK, Gillard JH, et al. (2002). Unilateral transplantation of human primary fetal tissue in four patients with Huntington's disease: NEST-UK safety report ISRCTN no 36485475. *Journal of Neurology, Neurosurgery, and Psychiatry* **73**:678–85.

39. Andrews PW, Damjanov I, Simon D, Banting GS, Carlin C, Dracopoli NC, Fogh J. (1984). Pluripotent embryonal carcinoma clones derived from the human teratocarcinoma cell line Tera-2. *Differentiation in vivo and in vitro. Laboratory Investigation* **50**:147–62.

40. Kondziolka D, Steinberg GK, Wechsler L, Meltzer CC, Elder E, Gebel J, Decesare S, Jovin T, Zafonte R, et al. (2005). Neurotransplantation for patients with subcortical motor stroke: A phase 2 randomized trial. *Journal of Neurosurg* **103**:38–45.

41. Kondziolka D, Wechsler L, Goldstein S, Meltzer C, Thulborn KR, Gebel J, Jannetta P, DeCesare S, Elder EM, et al. (2000). Transplantation of cultured human neuronal cells for patients with stroke. *Neurology 22* **55**:565–9.

42. Meltzer CC, Kondziolka D, Villemagne VL, Wechsler L, Goldstein S, Thulborn KR, Gebel J, Elder EM, DeCesare S, Jacobs A. (2001). Serial [18F] fluorodeoxyglucose positron emission tomography after human neuronal implantation for stroke. *Neurosurgery* **49**:586–91; discussion 91–2.

43. Nelson PT, Kondziolka D, Wechsler L, Goldstein S, Gebel J, DeCesare S, Elder EM, Zhang PJ, Jacobs A, et al. (2002). Clonal human (hNT) neuron grafts for stroke therapy: Neuropathology in a patient 27 months after implantation. *The American Journal of Pathology* **160**:1201–6.

44. Bang OY, Lee JS, Lee PH, Lee G. (2005). Autologous mesenchymal stem cell transplantation in stroke patients. *Annals of Neurology* **57**:874–82.

Part II

Practial/Methodological Chapters

Chapter 4

Differentiation of Neuroepithelia from Human Embryonic Stem Cells

Xiaofeng Xia and Su-Chun Zhang

Summary

We describe the method for efficiently differentiating human embryonic stem cells to neuroepithelial cells in a chemically defined condition. The protocol was established based on the fundamental principle of in vivo neuroectodermal development. The temporal course, morphological transformation, and shift in gene expression of our neuroepithelial differentiation closely resemble those occur during in vivo development. In particular, the primitive neuroepithelial cells generated by this protocol can be further induced into neuronal and glial cells with forebrain, mid/hind brain, and spinal cord identities and targeted transmitter phenotypes.

Key words: Human embryonic stem cells, Embryoid body, Neuroepithelia, Neuronal differentiation, Chemically defined condition

1. Introduction

Directed differentiation of human embryonic stem cells (hESCs) is key to dissecting early human development as well as producing lineage- and stage-specific cells for pharmaceutical screening and potentially cell therapy. To that end, we have devised a generic protocol for directing hESCs to synchronized neuroepithelial cells efficiently in a chemically defined system. This protocol is built upon the fundamental principle of neuroctodermal specification and the basic timeline of human embryo development. Technically, it avoids the harsh cell dissociation approach employed in mouse ESC differentiation technique which is not amenable for

Neil J. Scolding and David Gordon (eds.), *Methods in Molecular Biology, Neural Cell Transplantation vol. 549*
DOI: 10.1007/978-1-60327-931-4_4, © Humana Press, a part of Springer Science+Business Media, LLC 2009

hESCs. It also stays away from a co-culture with stroma cells to prevent biasing the differentiated cells to a particular regional progenitor (e.g., mid/hind brain progenitors) and being contaminated by carryover of tumorigenic stroma cells *(1)*. The protocol described below is chemically defined and typically yields over 90% of neuroepithelial cells among the total differentiated progenies, defined by immunostaining for the neuroepithelial transcription factors Pax6, Sox1, and Sox2 *(2)*. More importantly, this method allows control of developmental stages and generation of primitive neuroepithelial cells which can be further induced to neuronal and glial progenitors with forebrain, mid/hind brain, and spinal cord identities *(3, 4)*. Thus, this neuroepithelial differentiation method can be used as generic approach for generating neural progenitors and mature neural subtypes, as well as adapted to the needs of individual investigators who intend to differentiate hESCs to specific classes of neurons and glial cells *(5)*.

The protocol is a simplified and optimized version of a previous reported adherent colony culture *(6)*. It comprises three major steps: aggregation of hESCs ("embryoid body" formation), differentiation of multipotential primitive neuroepithelial cells, and generation of region-specific definitive neuroepithelial cells. Each step is morphologically distinct and is readily identifiable under a regular phase contrast microscope and typical photos have been provided as a guideline. The protocol has been followed by many amateur cell culture practitioners with consistent results. The key is that the hESC culture is free of partially differentiated cells.

2. Materials

2.1. Supplies

1. Polystyrene flask with polyethylene filter cap, T25 and T75 (Fisher Scientific, Pittsburgh, PA; cat. No. 12-565-57 and 12-565-31; or Nunc, Roskilde, Denmark; cat. No. 136196 and 178891).

2. Polystyrene plate, 6-well and 24-well (Fisher Scientific; cat. No. 12-565-73 and 12-565-75; or Nunc; cat. No. 140675 and 143982).

3. Polystyrene Petri dish, 60 mm (Fisher scientific; cat. No. 08-757-13A).

4. Polystyrene conical tube, 15 and 50 ml (Fisher scientific; cat. No. 05-527-90 and 14-432-23; or BD Bioscience, Bedford, MA; cat. No. 352095 and 352073).

2.2. Stock Solutions

1. *Dulbecco's modified Eagle's medium (DMEM)*. Nutrient mixture F-12 1:1 (DMEM/F12) (Gibco-BRL; cat. No. 11330-032).

2. L-Glutamine solution (200 mM) (Sigma, St. Louis, MO; cat. No. G7513). Make aliquots of 2.5 ml and store at –20°C.

3. MEM nonessential amino acids solution (Gibco-BRL, Rockville, MD; cat. No. 11140-050).

4. Knockout serum replacer (Gibco-BRL; cat. No. 10828-028). Make aliquots of 50 ml and store at –20°C.

5. Fetal bovine serum (Gibco-BRL; cat. No. 16000-044).

6. N2 supplement (Gibco-BRL; cat. No. 17502-048).

7. β-Mercaptoethanol (14.3 M) (Sigma; cat. No. M7522).

8. Recombinant human FGF basic (bFGF, R&D systems, Minneapolis, MN, cat. No. 233-FB) is dissolved in sterile PBS with 0.1% bovine serum albumin (Sigma; cat. No. A-7906) at a final concentration of 100 μg/ml. Make Aliquots of 50 μl and store at –80°C.

9. *Dispase solution (1 mg/ml)*. Dissolve 50 mg dispase (Gibco-BRL; cat. No. 17105-041) in 50 ml DMEM/F12 in a water bath for 15 min and filter-sterilize the dispase solution with a 50 ml steri-flip (Fisher Scientific; cat. No. SCGP00525).

10. *Heparin (1 mg/ml)*. Dissolve 10 mg heparin (Sigma; cat. No. H3149) in 10 ml DMEM/F12 medium. Make aliquots of 0.5 ml and store at –80°C.

11. Laminin from human placenta (Sigma; cat. No. L6274).

2.3. Media

1. The hESC growth medium. First add 3.5 μl β-mercaptoethanol to 2.5 ml of l-glutamine solution, then combine it with 392.5 ml DMEM/F12, 100 ml Knockout serum replacer and 5 ml MEM nonessential amino acids solution. Sterilize by filtering through a 500 ml filter unit (0.22 μm sterilizing low protein binding membrane) (Corning Incorporated, Corning, NY; cat. No. 430513). The medium can be stored at 4°C for up to 10 days. Add bFGF to final 4 ng/ml prior to use (*see* **Note 1**).

2. The neural induction medium. Sterilely combine: 489.5 ml DMEM/F12, 5 ml N2 supplement, 5 ml MEM nonessential amino acids solution, and 0.5 ml of 1 mg/ml heparin. Medium can be stored at 4°C for up to 2 weeks. bFGF (10 ng/ml) may be added prior to use.

3. Methods

3.1. Making ES Aggregates/ "Embryoid Bodies" (Day 1–4)

1. hESCs **(Fig. 1a)** are grown to confluent in a 6-well plate. Aspirate the ESC growth medium off and add 1 ml of Dispase to each well (The volumes used below are all for 1 well of a 6-well plate).

2. Check the cells every 2 min until the edges of cell colonies begin to curl off of the plate **(Fig. 1b)**. This usually takes 2–5 min when fresh dispase solution is used (*see* **Note 2**).

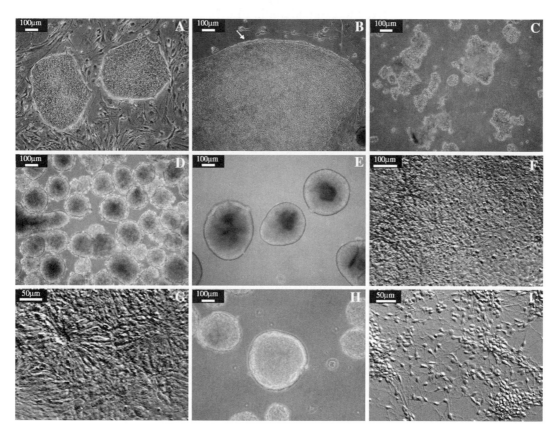

Fig. 1. Differentiation of neuroepithelia from hESCs. (**a**) hESC colonies. (**b**) A hESC colony was treated with dispase. The edge of the colony curled up as indicated by the arrow. (**c**) hESC were detached and pipetted into fragments of a couple of hundred μm in diameter. (**d**) hESC aggregated as spheres and floated in the medium. (**e**) After 3 days in the neural induction medium, cell aggregates became bright and clear. (**f**) Cell aggregates formed monolayer after attachment, "rosette" structures were observed about 4–5 days after attachment. (**g**) The columnar neuroepithelia cells formed neural tube-like rosettes. (**h**) Neuroepithelial cells were collected and grew as neuroepithelial clusters in suspension. (**i**) Neuroepithelial cells differentiated into neurons after attachment.

3. Add 3 ml hESC growth medium and gently detach the ESC colonies by pipetting using a 1,000 µl tip. Transfer the cells to a 15-ml conical tube. Gently triturate 3–5 times to break cell colonies into smaller clusters of about 100–200 µm in diameter using a 10-ml pipette (**Fig. 1c**).

4. Spin the hESC clusters at 200 × g for 2 min to settle them to the bottom of the tube. Aspirate off the medium.

5. Wash the cells once by adding 5-ml fresh hESC growth medium and then centrifuge for 2 min at 200 × g.

6. Aspirate off supernatant and resuspend cells in 5 ml of the hESC growth medium and transfer to a 60-mm Petri dish or a T25 flask (*see* **Note 3**).

7. Next day, undifferentiated hESCs will form aggregates and mostly float in the medium (**Fig. 1d**). Release those loosely attached aggregates by gently swirling the dish. If there are MEF and/or differentiated hESCs that attach to the dish, simply transfer the ESC aggregates to a new dish to remove the contaminated cells (*see* **Note 4**).

8. Feed the cells with the hESC growth medium every day for the first 4 days. When feeding, use a 5-ml pipette to gently pull aggregates up and then blow them back into the medium 2–3 times. This will help clean dead cells off the aggregate surface. Let the clusters settle to the bottom in a standing flask and aspirate off the medium.

3.2. Differentiating to Primitive and Definitive Neuroepithelia (Day 5–17)

1. Collect the hESC aggregates by centrifuging for 2 min at 200 × g and wash once with 5 ml of neural induction medium.

2. Resuspend cells in 5 ml of neural induction medium and transfer to a new T25 flask.

3. Feed the cells with neural induction medium every other day. Aggregates should become bright and clear after a few days in neural induction medium, healthy cell aggregates should be round and have smooth edges (**Fig. 1e**).

4. Prepare the laminin coated plates 2 days after cells in the neural induction medium (day 6). To coat, use 20 µg/ml mouse or human laminin in neural induction medium and leave the coated plates in 37°C incubator overnight (*see* **Note 5**).

5. Cells should be ready for attachment after 3 days in the neural induction medium. 30–50 aggregates are deposited in fresh neural induction medium in each well of a 6-well plate. Distribute the aggregates evenly over the plate to prevent contacting one another. This can be done by shaking the plate on the incubator shelf up and down twice and then left and right twice gently when placing the plate into the incubator (*see* **Note 6**).

6. Attached aggregates will collapse to form a monolayer colony after 1–2 days. Continue feeding with the neural induction medium every other day (*see* **Note 7**).

7. After 10–11 total days of differentiation (4–5 days following attachment), over 95% of the colonies should take on a morphology in which the center cells exhibit an elongated, columnar morphology (**Fig. 1f**). We call these columnar epithelial cells primitive neuroepithelial because they express a range of early neuroectodermal markers including Pax6, Sox2, and a host of anterior transcription factors. Cells at this stage are multipotential and can be used to differentiate to neuronal and glial types with distinct regional specificities.

8. Neuroepithelial cells should be fed with the same neural induction medium every other day and cultured for another 7 days. During this period starting at day 14–15 the columnar neuroepithelia cells will further proliferate, often forming ridges or rings of cells outlining a distinct lumen (**Fig. 1g**). The overall morphology is reminiscent of the neural tube and the structures at this stage are often referred to as neural tube-like rosettes.

9. After 17–18 days of differentiation under these conditions, the neuroepithelia that makes up the rosettes will stain positive for the definitive neural tube stage marker Sox1.

3.3. Isolating Definitive Neuroepithelia (Day 17–18)

1. Scratch off any colonies that do not contain any neuroepithelial cells (this should be less than 10% of colonies). The "bad" colonies can be marked by an objective marker lens under a phase contrast scope and then scraped away with a pipette tip in a sterile hood.

2. Dislodge the neuroepithelia by gentle pipetting with a 1,000-μl tip. Neuroepithelial cells are denser in the center of colonies so that they are easily detached as opposed to the flat, tightly bound cells at the periphery.

3. Collect the detached neuroepithelial cells in a 15-ml conical tube and spin them down at $100 \times g$ for 2 min. Wash the cells once with fresh neural induction medium.

4. Aspirate off the medium and resuspend the clusters of neuroepithelia in 5 ml of neural induction medium, supplement with B27 to improve cell survival. Transfer the cells to a new T25 flask.

5. Over the next 24 h the rosette aggregates will roll up to form round spheres (**Fig. 1h**), while any flat nonrosette peripheral cells will usually attach to the culture vessel. After this period, rosette aggregates should be switched to a new flask and grown in neural induction medium (*see* **Note 8**).

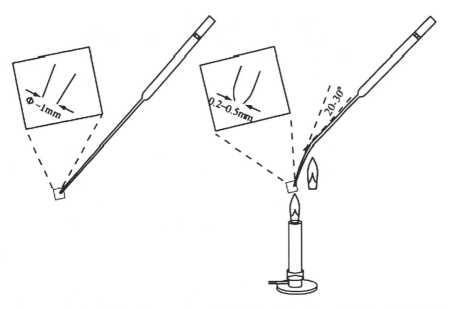

Fig. 2. Pasteur pipette technique for breaking up neuroepithelial clusters. Flame the end of cotton-filtered Pasteur pipette to round the edges and also narrow the opening to about 0.2–0.5 mm in diameter. Flame the narrow part of the shaft to introduce a 20–30° bend. Rinse the pipette wall with the culture medium three times. Pipette neuroepithelial clusters using the treated Pasteur pipette by pull the cells in and push them out for three times. The narrow opening and bend in the pipette will help shear the clusters into smaller pieces of roughly uniform size.

Neuroepithelial cells can be maintained in the neural induction medium until they are used for further transplantation purpose or in vitro differentiation to neuronal and glial cells (**Fig. 1i**). Aggregates will stop growing when a maximal size is reached. To help cell proliferation, aggregates should be broken down to smaller sizes using the Pasteur pipette technique described in **Fig. 2**. In this way, the total number of cells can be amplified and neuroepithelia can be propagated for several months in the neural induction medium with bFGF or other growth factors. However, the types of neural cells generated will inevitably change with long-term culture.

4. Notes

1. Recombinant growth factors lose activities quickly in diluted condition. Do not add them into the medium until use.
2. If the colonies do not start to curl off in 5 min, old dispase solution should be discarded and change to a fresh solution.

3. Before placing the dish or flash into the incubator, shake to disperse the cell clumps to avoid aggregation.

4. Do not pipette to release the flat attached cells. They are residual MEFs or differentiated cells that should be avoided in the following steps.

5. We have found 3 h of coating to be sufficient, coating for less than 3 h is not recommended.

6. Do not swirl the plate. Swirling will almost inevitably lead to aggregation in the center of the well. Cell differentiation will be affected due to lack of space in between. It often works best to set aggregates up for attachment at the end of the day to minimize traffic in and out of the incubator. As long as aggregates are relatively bright and clear after 6 days in culture they should attach to the dish over the 24–48-h period. If aggregates do not look healthy they can be kept in neural induction media for an additional 1–2 days. Washing cells several times with neural induction media will help remove all the loose cells from the aggregate surface which can inhibit attachment. An alternate attachment method is to supplement 10% fetal bovine serum in the neural induction medium for 12–24 h. The serum should be removed and washed after 24 h as it will inhibit neural induction. It should also be avoided if it interferes with the aim of the experiment.

7. Cells attach loosely in the first 1–2 days. Be gentle and not to detach them when changing medium.

8. The rosette aggregates are sometimes called neurospheres although they are quite different from the neurospheres prepared from brain tissues.

References

1. Du, Z.W. and Zhang, S.C. (2004). Neural differentiation from embryonic stem cells: which way? *Stem Cells Dev.* **13**, 372–381.

2. Pankratz, M.T., Li, X.J., Lavaute, T.M., Lyons, E.A., Chen, X., Zhang, S.C. (2007). Directed Neural Differentiation of hESCs via an Obligated Primitive Anterior Stage. *Stem Cells.* **25**, 1511–1520.

3. Li, X.J., Du, Z.W., Zarnowska, E.D., Pankratz, M., Hansen, L.O., Pearce, R.A., Zhang, S.C. (2005). Specification of motoneurons from human embryonic stem cells. *Nat Biotechnol.* **23**, 215–221.

4. Yan, Y., Yang, D., Zarnowska, E.D., Du, Z., Werbel, B., Valliere, C., Pearce, R.A., Thomson, J.A., Zhang, S.C. (2005). Directed differentiation of dopaminergic neuronal subtypes from human embryonic stem cells. *Stem Cells.* **23**, 781–790.

5. Zhang, S.C. (2006). Neural subtype specification from embryonic stem cells. *Brain Pathol.* **16**, 132–142.

6. Zhang, S.C., Wernig, M., Duncan, I.D., Brustle, O., Thomson, J.A. (2001). In vitro differentiation of transplantable neural precursors from human embryonic stem cells. *Nat Biotechnol.* **19**, 1129–1133.

Chapter 5

Derivation of High-Purity Oligodendroglial Progenitors

Maya N. Hatch, Gabriel Nistor, and Hans S. Keirstead

Summary

Oligodendrocytes are a type of glial cells that play a critical role in supporting the central nervous system (CNS), in particular insulating axons within the CNS by wrapping them with a myelin sheath, thereby enabling saltatory conduction. They are lost, and myelin damaged – demyelination – in a wide variety of neurological disorders. Replacing depleted cell types within demyelinated areas, however, has been shown experimentally to achieve remyelination and so help restore function. One method to produce oligodendrocytes for cellular replacement therapies is through the use of progenitor or stem cells. The ability to differentiate progenitor or stem cells into high-purity fates not only permits the generation of specific cells for transplantation therapies, but also provides powerful tools for studying cellular mechanisms of development. This chapter outlines methods of generating high-purity OPCs from multipotent neonatal progenitor or human embryonic stem cells.

Key words: High-purity oligodendrocyte precursor cultures, Oligodendrocytes, Remyelination, Oligodendrocyte precursors

1. Introduction

Oligodendrocytes are a type of glial cells that play a critical role in supporting the central nervous system (CNS). They insulate axons within the CNS by wrapping them with a myelin sheath, thereby enabling saltatory conduction. Demyelination refers to the loss of the myelin sheath, and characterizes a number of disorders including but not limited to spinal cord injury, multiple sclerosis, brain trauma, stroke, Alzheimer's disease, adrenoleukodystrophy, Guillain–Barre syndrome, Canavan disease, post-infectious encephalitis, and chronic inflammatory polyneuropathies (1–9). In addition to a loss of saltatory conduction in axons, demyelination

Neil J. Scolding and David Gordon (eds.), *Methods in Molecular Biology, Neural Cell Transplantation, vol. 549*
DOI: 10.1007/978-1-60327-931-4_5, © Humana Press, a part of Springer Science+Business Media, LLC 2009

can also lead to axon degeneration and further disease progression *(7, 10–12)* that cannot be rectified by the limited endogenous repair mechanisms *(13–19)*.

Remyelination refers to replacement of the myelin sheath, which can restore saltatory conduction *(20, 21)*, improve locomotor deficit *(22–24)*, and prevent further axonal loss and disease progression by improving axonal integrity *(23, 25, 26)*. Remyelination can be achieved by replacing depleted cell types within demyelinated areas, and has been demonstrated using oligodendrocyte precursor cells (OPCs), Schwann cells, and olfactory ensheathing glia *(27–33)*.

One method to produce OPCs for cellular replacement therapies is through the use of progenitor or stem cells. These OPC sources have been used in several animal models of demyelination, and have been shown to integrate, differentiate, and remyelinate damaged CNS *(23, 34–39)*. Derivation of high-purity OPC cultures is one of the major challenges facing the development of therapeutic strategies to address demyelination. This chapter outlines methods of generating high-purity OPCs from multipotent neonatal progenitor or human embryonic stem cells (hESCs). The ability to differentiate progenitor or stem cells into high-purity fates not only permits the generation of specific cells for transplantation therapies, but also provides powerful tools for studying cellular mechanisms of development.

2. Materials

2.1. OPC Generation from Neural Progenitor Cells

2.1.1. Supplies

1. Low adherent 6-well plates (Corning, Cat #3471).
2. 50-ml centrifuge tubes (Fisher #: 05-538-68).
3. 1 l Vacuum Filters 0.22 μM polystyrene (Corning #: 431205).
4. Permanox Lab-Tek™ Chamber Slides™ (Fisher #: 12-565-21).
5. T75/75 cm² flasks (BD Biosciences #: 13-680-65).
6. Trypan blue (Sigma #: T8154).

2.1.2. Solutions

1. DMEM:F12 media (Gibco #: 10565-018).
2. DMEM media (Gibco #: 12430-047).
3. 0.05% trypsin–EDTA (Gibco #: 25300-054).
4. Antitrypsin (Sigma #: T7659).
5. Poly-L-lysine (Sigma #: P4707).
   – Laminin (Sigma #: L2020.

2.1.3. Recipes

Oligodendrocyte Culture
Media

1. 1,000 ml of DMEM:F12 media.
2. 2 × 10 ml of 1:50 B27 supplement.
3. Insulin (10 μg/ml).
4. Progesterone (63 ng/ml).
5. Putrescine (10 μg/ml).
6. Sodium selenite (50 ng/ml).
7. Transferrin (50 μg/ml).
8. T3 (40 ng/ml).

Filter through 0.22 μm pore

2.2. OPC Generation from Human Embryonic Stem Cells

2.2.1. Supplies

1. 15 ml conicals (Fisher #: 14-959-11B) and 50 ml conicals (Fisher #: 05-538-68).
2. 1.5 ml cppcndorf tubcs (Fishcr Scientific #: 05 402 96).
3. BD biosciences T75 and T150 cell culture flasks (Fisher #: 13-680-65 and 08-772-48).
4. 6-well Costar low attachment plates (Corning #: 3471).
5. *Barrier tips.* 200 μl and 1,000 μl (any vendor).
6. 1 l Vacuum Filters.22 μM polystyrene (Corning #: 431205).
7. Permanox Lab-Tek™ Chamber Slides™ (Nunc #: 177437).
8. Trypan blue (Sigma #: T8154).

*2.2.2. Stock Media/
Solutions*

1. 70% ethanol.
2. 30% Bleach.
3. DMEM:F12 media (Gibco #: 10565-018).
4. KO-DMEM media (Gibco #: 10829-018).
5. DMEM media (Gibco #: 12430-047).
6. Water for embryo transfer (WET) (Sigma catalog #W1503).
7. D-PBS (Gibco catalog #14190-144).
8. 0.05% Trypsin–EDTA (Gibco #: 25300-054).
9. Trypsin inhibitor or antitrypsin (Sigma #: T7659).
10. Bovine serum albumin (Sigma #: A2153).

2.2.3. Supplements

See **Table 1**. Supplements are prepared according to the manufacturer specification and stored in 1 ml aliquots at –20°C at 1,000× the stock concentration.

2.2.4. Recipes

1. Mouse embryonic feeder conditioned medium (MEF-CM) – see published protocols.
2. Glial restrictive media (GRM) for 1 l:
   – 1,000 ml of DMEM:F12 media.

Table 1
Required supplements

Component	Manufacturer (catalogue #)	Stock concentration	Final Concentration	Dilution media
B27	Gibco (17504-044)	50×	1:50	n/a
Insulin	Sigma (I-1882)	10 mg/ml	10 μg/ml	100 μl glacial acetic acid, rest w/WET
Progesterone	Sigma (P-6149)	63 μg/ml	63 ng/ml	1 ml of EtOH then rest w/KO-DMEM
Putrescine	Sigma (P-6024)	10 mg/ml	10 μg/ml	KO_DMEM
Sodium selenite	Sigma (S-9133)	50 μg/ml	50 ng/ml	WET
Transferrin	Sigma (T-1408)	50 mg/ml	50 μg/ml	KO-DMEM
T3	Sigma (T-5516)	40 μg/ml	40 ng/ml	Add 1 ml of 1 N NaOH then KO-DMEM
Collagenase IV	Gibco (17101-015)	1 mg/ml	1 mg/ml	–
bFGF	Gibco (13256-029)	10 μg	4 ng/ml	0.5% BSA/PBS
RA[a]	Sigma (R-2625)	20 mM	10 μM	DMSO
EGF	Sigma (E-9644)	20 μg/ml	20 ng/ml	KO-DMEM

[a]Add 1 ml of DMSO to the vial and swirl to dissolve. Transfer the solution to a 15 ml tube and wash the ampule with 200 μl of DMSO. Add DMSO to the final volume of 8.3 ml. Make 100–300 μl aliquots of this mixture in light protected vials and store at –80°C

- 2 × 10 ml of 1:50 B27 supplement.
- 1 vial of stock insulin (10 μg/ml).
- 1 vial of stock progesterone (63 ng/ml).
- 1 vial of putrescine (10 μg/ml).
- 1 vial of sodium selenite (50ng/ml).
- 1 vial of transferrin (50 μg/ml).
- 1 vial of T3 (40 ng/ml).
- Mix the components in a filter cup/bottle and vacuum filter. Homogenize by gently swirling the bottle. Avoid media on the bottle cap. Warm up only the required volume.

3. Transition medium (TR).
   - Made as a 1:1 mixture of GRM and MEF-CM.

Other solutions:

4. Matrigel coating 1:30. Matrigel will jellify instantly at room temperature. Manipulations after thawing have to be done quickly in precooled recipients on ice.

- Obtain growth factor reduced Matrigel from BD Biosciences (Cat #356231). Place the bottle in a cold water bath and before complete thaw, open the cap and add 10 ml KO-DMEM. Dissolve the remaining frozen block by pipetting up and down (avoid foam formation) and quickly aliquot into 15 ml centrifuge tubes, 2 ml of 1:2 Matrigel in each. These aliquots have to be immediately frozen and kept at –20°C until needed.

- To prepare the working solution, 8 ml of KO-DMEM is added to each 2 ml frozen aliquot for a total volume of 10 ml. For the final dilution, an additional 20 ml of KO-DMEM is added to each aliquot and then distributed into the flasks (for example 10 ml in each). Gently swirl the coating solution in the flask and place it in a 4°C refrigerator. The flasks are coated in approximately 3 h and can be used the same or next day.

- One hour before use, the coating Matrigel solution is discarded from the flask and replaced with the working media formulation. The flask is placed in a CO_2 incubator for pH and temperature balance.

2.3. Markers and Dilutions

See Table 2

Table 2
Common antibodies used for ICC profiling to identify OPC cultures are listed in the table with their respective dilutions

Antibody	Host	Dilution
A2B5	Mouse	1:200
Pax 6	Rabbit	1:200
Olig1	Rabbit	1:200
Sox10	Rabbit	1:200
Nestin	Mouse	1:200
NG2	Rabbit	1:200
O4	Mouse	1:50
PDGFaR	Mouse	1:100
SSEA-4	Mouse	1:100
Tuj1	Mouse	1:500
GFAP	Rabbit	1:500

3. Methods

3.1. OPC Generation from Neural Progenitor Cells

3.1.1. Isolating and Plating Neural Progenitor Cells

Day 1: CNS neural progenitor cells are isolated from four to five pups

1. Prepare oligodendrocyte culture media (see recipe below).

2. Obtain and place 1-day old newborn pups on ice until comatose. Spray with EtOH and decapitate with scissors. While holding the head by the nose, carefully cut the skin and skull down the midline from back to front and peel away.

3. Lift the brain from the olfactory bulbs and place in a Petri dish filled with Hank's balanced salt solution supplemented with penicillin and streptomycin on ice. Remove meninges and major blood vessels then cut away the frontal lobes, cerebellum, and brainstem. Cut away the cortices and isolate the striatum. Once isolated, place the tissue in another Petri dish containing cold culture media.

4. Repeat above steps with the rest of the pups until all of the striata are isolated and in the same Petri dish.

5. Aspirate media and mince tissue with sterile scalpel blade.

6. Add 10 ml of 0.05% trypsin–EDTA, incubate at 37°C for 5 min, gently triturate with fire-polished Pasteur pipette, and then return to incubator for another 5 min.

7. Add equal amounts 0.05% antitrypsin, spin down cells at 400 × g for 5 min, aspirate supernatant, and resuspend in 30 ml warm media supplemented with 20 ng/ml EGF. Transfer cellular suspension to uncoated low-adherent 6-well plates dispensing 5 ml per well. Place in incubator at 37°C and 5% CO_2 (*see* **Note 1**).

3.1.2. Growing and Feeding Spheres

Day 2: Debris is removed at this stage and cells will agglomerate in clumps resembling neurospheres

1. Warm 30 ml of media.

2. Collect cells from 6-well plates and put into a 50-ml conical tube.

3. Spin down cells at 400 × g for 5 min. Aspirate supernatant and resuspend in fresh media supplemented with 20 ng/ml of EGF. Distribute 5 ml per well in the 6-well plate and return to incubator.

Day 3: Prepare plates for plating

1. Coat T75 (75 cm²) flask using 10 mg/ml poly-l-lysine and 15 μg/ml laminin making sure that the surface is well coated. Leave at 4°C overnight. Before use, aspirate excess coating mixture, rinse one time with PBS, and replace with culture media.

3.1.3. Plating Neuro-
spheres

Day 4: Neurospheres are plated, allowing for cell migration

1. Warm media.

2. Collect cells and place into a 50-ml conical tube. Spin cells at $400 \times g$ for 5 min, aspirate supernatant and resuspend in fresh media without EGF. Add spheres to coated T75 and place in incubator. Cells will adhere overnight and begin to migrate.

3.1.4. Day 5–6:
Transplantation

1. Observe the cells at day 5 and if growing well (i.e. migrated and relatively confluent), then they are ready for use. If growing slower, change the media and keep for the next day.

2. For transplantation:

   (a) Add a small amount of 0.05% trypsin–EDTA to flask and incubate at 37°C for 5 min. Take cells out of the incubator, add an equal amount of antitrypsin and triturate by aspirating up and down with a serological pipette to break apart large clusters.

   (b) Spin at $400 \times g$ for 5 min, aspirate supernatant and wash/resuspend cells with 10 ml of DMEM. Take a small sample of cells to perform cell count. Use 50 μl of cells with 50 μl of trypan blue. Perform cell count to determine total number of cells while washing cells two more times with media.

   (c) On the last wash, resuspend cells in the amount of transplanted vehicle (DMEM or Hanks Balanced Salt Solution) needed to give you a final concentration of 12×10^4 cells/μl.

3. For in vitro characterization:

   (a) Prepare polylysine–laminin coated slides using Lab tek chamber slides 24 h in advance. Prepare cells as described for transplantation. Resuspend cells in oligodendroycte culture media. Plate to a final concentration of 150,000 cells/cm². To assess differentiation potential, cells can be grown on plated imaging slides for 7 days before fixation with 4% paraformaldehyde.

3.2. OPC Generation from Human Embryonic Stem Cells

3.2.1. Growing Stem Cells

hESCs are grown and expanded on Matrigel-coated flasks until the appropriate numbers for differentiation are achieved.
Day before plating hESCs:

1. Coat T75 flasks by adding 1:30 diluted growth factor reduced Matrigel in KO-DMEM to flask. Swirl flask to ensure even spreading of solution and set in a 4°C refrigerator overnight. Before use, aspirate the Matrigel solution, replace with 20 ml of MEF-CM and place the flask for at least 1 h in the incubator for thermal and pH balance.
Day 1: Plating and initial growth

2. Warm 35 ml of MEF-CM in a 37°C water bath.

3. Quickly thaw a 1 ml vial of hESCs with approximately 1.5×10^6 cells and add 9 ml of MEF-CM. Spin cells down at $200 \times g$ for 4 min and resuspended in 5 ml of MEF-CM media. Carefully break up the pellet. Do not break the cell clusters.

4. Add the resuspended cells to the flask and add 8 ng/ml of bFGF. Place the flask in an incubator at 37°C, 5% CO_2.

Day 7–14: Growth and splitting/passaging.

5. Cells are fed with 20–30 ml of new MEF-CM and 8 ng/ml bFGF everyday. MEF-CM media is warmed and old media is aspirated and replaced with new MEF-CM + 8 ng/ml bFGF.

6. Once a week cells are split 1:4 or 1:6 when at a minimum of 70% confluence.

7. Coat plates with Matrigel as described above, 24 h in advance

8. Prepare MEF-CM + 8 ng/ml bFGF

9. Dissociate cells from the flasks by adding 10 ml of 1 mg/ml collagenase IV to the flask and leave in the incubator for 2–5 min (*see* **Note 2**).

10. Aspirate collagenase and wash cells with 10 ml D-PBS. Aspirate D-PBS and add 10 ml of new MEF-CM to the flask. Scrape the hESCs with a cell scraper.

11. Collect the cells and distribute according to the splitting ratio into newly coated flasks, add 8 ng/ml of bFGF to each.

12. Replace flasks in incubator.

13. Feed the cells daily by replacing with fresh media and 8 ng/ml of bFGF.

3.2.2. Cellular Aggregates (Start of Differentiation)

Differentiation is initiated by removing the cells from the adherent substrate and treating them with a new media composition (transition media) for the first 2 days and continuing with GRM for the rest of the protocol duration – to induce neural differentiation first, and later oligodendroglial differentiation.

Day 1:

1. Prepare 30 ml of transition media (TR) + 4 ng/ml bFGF for each T75 flask, prewarm and pH balance in a CO_2 incubator.

2. Treat hESCs with 10 ml per flask of 1 mg/ml collagenase IV for 2–5 min in incubator. Aspirate collagenase and wash cells with 10 ml of D-PBS.

3. Aspirate D-PBS and add 30 ml of new TR media to cells then scrape cells with cell scraper to dislodge. Collect cells from the flask into a 50 ml centrifuge tube and triturate cell aggregates to slightly break up large clumps (about 3–5 times).

4. Distribute the cells to Costar low attachment 6-well plates (5 ml cell suspension in each well). Put in incubator at 37°C, 5% CO_2.

Day 2:

5. Prepare 30ml of TR media + 4 ng/ml bFGF + 20 ng/ml EGF and 10 μM retinoic acid (RA) for each 6-well plate of cells (*see* **Note 4**).

6. Collect cells from each well in 6-well plates and combine into 50 ml centrifuge tube. Spin cells at 200 × g for 4 min.

7. Aspirate old media and add 30 ml of new TR media with supplements. Resuspend gently, and do not break the clusters.

8. Distribute 5 ml to each well in plate. Return cells to incubator.

3.2.3. Yellow Sphere Formation/Neural Progenitor

Day 3–10: Small clusters of 20–50 cells can be seen floating in the media. At the end of this time period, yellow spheres can be clearly observed as the only growing elements in the culture. The use of RA everyday will cause caudalization of the progenitor population (*see* **Note 3**).

1. Make GRM media and warm. Prepare 30–35 ml of GRM + 20 ng/ml of EGF + 10 μM of RA for each 6-well plate of cells. Use minimal light during feeding since RA is light sensitive. For the RA, use 1 μl for each ml of media from the 10 μM/ml stock solution. Keep the RA stock solution aliquoted according to the experiment size, in 100, 200, 300 μl cryovials, at –80°C. The thawed vial should be discarded after use.

2. Collect cells in 50-ml conical tube and centrifuge at 200 × *g* for 2 min. If supernatant contains debris, eliminate with a low force centrifugation.

3. Aspirate supernatant and add 30 ml of new GRM + EGF + RA media.

4. Do not dissociate clumps by aspiration.

5. Redistribute 5 ml in each well.

6. The procedure is done every day.

Day 11–15: Medium yellow clusters/spheres are the only cellular aggregates growing in the culture, and very likely to be macroscopically visible. The dark clusters have an irregular shape and a loose composition. Individual floating cells are discarded at every feeding procedure.

7. The cultures are fed every other day (Monday–Wednesday–Friday).

8. Prepare 30 ml of GRM media + 20 ng/ml EGF for each plate of hESCs.

9. Collect cells and perform the same procedures as described for days 3–10 with a low centrifugation force (200 × *g* for 1–2 min)

Day 16–28: Large clusters

Yellow sphere growth continues as unhealthy cells are discarded.

10. Change media every M–W–F and always return cells to the incubator as soon as possible.

11. Prepare 30 ml of GRM media + 20 ng/ml of EGF for each plate of cells.

12. Cells are collected in 50 ml tubes and are left to settle on their own for 5–10 min.

13. Aspirate supernatant and add new media.

14. Agitate the tube and immediately collect 15 ml using a 25 ml pipette and distribute quickly in first three wells.

15. Agitate the tube again and collect the remaining 15 ml and distribute the rest in the other wells.

3.2.4. Oligodendrocyte Progenitors

Day 28: Purification is obtained by plating the yellow spheres on Matrigel-coated flasks. Plating will eliminate the dead or non-adherent cells and will promote an outward migration from the yellow spheres.

1. Prepare Matrigel-coated T75 flask with 1:30 Matrigel in KO-DMEM 24 h in advance.

2. Prepare 30 ml of GRM + 20 ng/ml EGF, place the media in the coated flask, and prebalance the temperature and pH for at least 1 h in the CO_2 incubator.

3. Collect cells from flask into a 50 ml centrifuge tube. Let the culture sit for 5 min.

4. Aspirate old media and add a small amount (5 ml) of pre-warmed GRM + EGF.

5. Place the cells into the coated T75 flask with the balanced GRM.

6. Return to incubator at 37°C, 5% CO_2.

7. The next day, gently shake the plated flask to dislodge non-adherent debris.

Day 29–34: Oligodendrocyte progenitors grow and migrate out from the yellow spheres.

8. Change media every M-W-F.

9. Prepare 30 ml of GRM + 20 ng/ml EGF for each Matrigel-coated flask.

10. Aspirate old media.

11. Add 30 ml of new GRM + EGF to flask. Return to incubator.

Day 35–42: Purification, replating and imaging

The cultures will go through a purification process where contaminant cells (astrocytes, fibroblasts) are removed and the remaining oligodendrocyte progenitors are replated. At the same time cells are plated on imaging slides, to be used later for immunocytochemistry, while the rest of the culture is prepared for in vivo use. A 1:2 splitting ratio is used for this step.

12. Prepare the required Matrigel-coated T75 flasks with 1:30 Matrigel in KO-DMEM 24 h in advance.

13. Add 25 ml of GRM + EGF to each flask; prewarm and balance the pH of the media in the CO_2 incubator 1 h before use.

14. Prepare Matrigel 1:30 or laminin (10 µg/ml/cm²) coated imaging slides one day in advance. Replace the coating solution with media, omitting the growth factor, and place the slides in the CO_2 incubator for temperature and pH balance 1 h before use.

15. Aspirate media from each cell containing flask. Wash with 10 ml D-PBS. Aspirate again.

16. Add 7 ml of warm 0.05% trypsin–EDTA to flask. Incubate 5–10 min at 37°C.

17. Add 7 ml of antitrypsin solution to the flask. Collect the cells into a 15 ml centrifuge tube and mix.

18. Take a small sample of cells to perform a cell count. Take 50 µl of cells and mix with 50 µl of trypan blue. Perform cell count for total number of cells recovered.

19. Spin the cells at $250 \times g$ for 5 min.

20. Aspirate media and resuspend in 50 ml of GRM media. Transfer cells to two T75 or one T150 bare plastic flasks and incubate for 1 h at 37°C. This step allows all of the astrocytes to attach to the plastic bottom, while the less adherent oligodendrocytes will float in the media.

21. Collect media with a gentle shake from the plastic flasks and transfer to centrifuge tubes.

22. Take a sample from the purified cells for the laminin imaging slides. Dilute cells to 50,000 cells/ml/cm². EGF is omitted from the media. After 2 days the cells are ready to be fixed and stained.

23. The rest of the purified cells are split into two Matrigel-coated T75 flasks prepared earlier containing GRM + 20 ng/ml EGF. After 7 days of growth, they are ready to be used (*see* **Note 5**).

3.3. Cell Identification

3.3.1. Morphology

Growing Stem Cells (Undifferentiated)

Every stage during hESC differentiation has distinct cellular or cellular aggregate morphology detailed below. If growing OPCs from striatal preparations, cellular descriptions start at sphere formation and their morphology is very similar to the hESC lines.

Healthy undifferentiated stem cells grow in circular clusters with a smooth looking surface. Some of the clusters will have some multilayered areas and some spontaneously differentiated areas. Individual cells will be in-between clusters as they spontaneously differentiate and migrate. Confluence is usually attained within 5–7 days.

Cellular Aggregates

Aggregate clusters appear after dissociation of hESCs cultures. The large stem cell clusters first break apart and then round up. They will not be perfectly spherical and they will have various sizes. Isolated floating cells are usually nonviable and are discarded at feeding with low force centrifugation. Some published research refers to these aggregates as "embryoid bodies". We have not demonstrated the presence of the three embryonic layers, so refer to them as "cellular aggregates" instead.

Yellow Spheres/Neural Progenitors

Spheres grow from the aggregate clusters, take a perfect spherical morphology, and are bright yellow. Small and large yellow spheres will be surrounded by cellular debris. The spheres should be homogenous and no dark, necrotic centers should be observed. This stage is crucial for selecting healthy spheres and patience is necessary for proper selection. Some spheres will adhere to each other.

During this stage, more and more yellow spheres are produced and there are fewer contaminants seen with time.

Oligodendrocyte Progenitors

Plating of neural progenitors allows selection of viable cells, self-dissociation of the spheres, and further differentiation of neural progenitors into OPCs. During this stage, the migrated cells can look flat reassembling an epithelial morphology or they can have a bipolar morphology with short thick branches. Most importantly, they are positive for oligodendrocyte markers such as Olig1 and NG2. Some plated yellow spheres will extend long processes first, followed by migration of the OPCs along the radial branches over the next few days.

Oligodendrocytes

After plating at low density in a growth factor free medium, some of the cells will develop the fully mature shape of oligodendrocytes with branches and sheets, staining positive for all oligodendroglial markers.

3.3.2. Markers

Note: These markers are generally good for both neural and embryonic derived populations.

Growing Stem Cells (Undifferentiated)

Good markers for this stage are:
 – SSEA-4: a surface marker of pluripotent cells
 – Oct4: nuclear marker, transcription factor

Cellular Aggregates

This is a transition stage and when plated on adherent substrate many cells will stain positive for embryonic markers (SSEA, OCT4) and while others will start to stain for neural progenitors such as nestin and A2B5 after approximately 3 days. They reassemble neurogenic cores surrounded by nonlabeled cells.

Spheres/Neural Progenitors

Early yellow spheres (up to day 21) plated on adherent substrate will stain positive for:

- Pax6: A transcription factor for neural commitment

- A2B5: marker for early neural progenitors

- Nestin: marker for neural commitment

- Olig1 or Olig2: a transcription factor involved in oligodendroglial development

Oligodendrocyte Progenitors

Typical markers for OPCs are:

- Olig1 or Olig2: transcription factor specific for pMN domain

- NG2: Chondrotin sulfate proteoglycan expressed by OPCs

- PDGFαR: specific for young oligodendrocytes or early OPCs

- SOX10: a shuttle protein (nuclear and cytoplasmic) involved in oligodendrocyte differentiation also positive in young and mature oligodendrocytes

Contaminants:

- GFAP: marker for astrocytes which contaminate the OPC cultures (usually less than 5%).

- Neurofilaments (bTubulin, MAP2, Tuj1 etc): positive for neurons (sometimes up to 10%)

- SMA (Smooth Muscle Actin): occasionally single cells are found in the culture (less than 0.1%)

Each of these antibodies requires slightly different protocols for immunostaining; we have provided a table above noting some of the differences. The general protocol for immunocytochemistry is provided below.

Immunostaining Procedure

1. Fix cultures for 10 min at room temperature (RT) in 4% paraformaldehyde

2. Wash 3× saline phosphate buffer (PBS) for 5 min each

3. Wash in 1% BSA + 0.1% trition-X in PBS for 15 min at RT

4. Wash 3× PBS for 5 min each

5. Dilute primary antibodies in 1% BSA in PBS at dilutions noted in the table above. Leave antibody incubation overnight in 4°C

6. Wash 3× PBS for 5 min each

7. Block with 10% goat serum in PBS for 30 min at RT

8. Wash 3× PBS for 5 min each

9. Add AlexaFluor secondary antibody according to the primary host species (antimouse or antirabbit) 1:200 in PBS for 1 h RT

10. Wash 3× PBS for 5 min each

11. Counter stain with Hoechst (1:1,000) for 5 min RT

12. Wash 3× PBS for 5 min each and one time with dH$_2$O

13. Coverslip with an aqueous mounting media

With Alexa Fluor conjugated secondary antibodies, we found persistent fluorescence after 1 year while the slides were kept at room temperature.

4. Notes

4.1. General Tissue Culturing Notes

Variations from the given protocol can be done with certain limitations. We found that the timing and the growth factor sequence is critical. The cultures are tolerant to changes in the concentration of supplements; however, the outcome will be reflected in the oligodendrocyte vs. contaminant cell yield.

1. *Contamination*. Due to the long protocol (42 days), most of the problems that arise in stem cell differentiation are a result of improper sterile technique resulting in contamination. It is best to always be as sterile as possible and be on the lookout for unusual growth. If contamination is caught early, the cells may still be salvaged. Repeated washing with sterile HBSS at feeding and addition of antibiotic for a week can restore the cultures. If contamination persists, the cultures must be destroyed and the incubator decontaminated. In such an event, we prefer to decontaminate the entire cell culture room. Some preventative procedures to follow are: (a) Always work in a certified biosafety hood and calibrate the incubators at least once a year. (b) Spray anything that goes into the hood with generous amounts of 70% ethanol or alternative disinfectant (bleach, or commercially available products). (c) Spray down the hood itself before and after each use. (d) Ethanol treat your gloves often. (e) Use barrier tips for all pipetting. (f) Do not leave wrappers from tips, used tubes, or extra debris in the hood. Always remove things directly after use. (g) Never leave the incubator door open. (h) Rinse aspiration tube after each use with 30% bleach. (i) Keep all flask caps and 6-well plate lids clean from media or liquid. Replace wet caps and lids with new, clean ones. (j) Keep all areas separate from other tissue/cellular use. If possible use the room and hood only for stem cells. This will decrease contamination from other cell lines. (k) Aliquot reagents or supplements into one dose vials. (l) Do not work with the cells when you are sick (flu, cold) and always wear a mask during the epidemic season. (m) As the major source of contamination, water baths should be completely eliminated from the critical cell culture rooms.

The media has to be left in the 37°C for 1 h for pH and temperature balance. (n) Check the water level in the incubator to prevent evaporation of media resulting in hyperosmolarity and sudden cell death.

2. *Extensive cellular death* at the stem cell culture dissociation step: This is caused by the overgrowth of the stem cell culture and cannot be recovered. Consider adding more media; make 2 × 6 well plates from one stem cell flasks.

3. *Yellow spheres are not forming.* Low density or nondifferentiating stem cells. The quality of the stem cell cultures is crucial for good differentiation and can be appreciated by the spontaneous differentiation tendency. A good starting stem cell culture should have abundant stromal cells, and large stem cell clusters sharply delimited with a smooth edge. In our experience, cultures with low or absent spontaneous differentiation (reflected in the abundance of stromal cells) generated very low yields of OPCs. In these cultures, hollow cystic cell agglomerates are common, and fail to evolve. Cultures with too much spontaneous differentiation (multilayered clusters) generated more neuronal contaminants.

4. *The retinoic acid quality* can be incriminated if failure of differentiation is observed. The old stock should be replaced after 3 months. A good, active RA is bright yellow, but is degraded very quickly by fluorescent light and the yellow intensity is visibly faded.

5. *Failing terminal differentiation.* The plating density for terminal differentiation in the absence of growth factors is critical. If the density is high, the cells will proliferate. If density is too low, the cells do not survive. During characterization of plated imaging slides, we often see a mixture of young and semimature oligodendrocytes.

References

1. Cervos-Navarro, J. and J. V. Lafuente (1991). "Traumatic brain injuries: structural changes." *J Neurol Sci* **103**(Suppl): S3–14.

2. Lassmann, H. (2001). "Classification of demyelinating diseases at the interface between etiology and pathogenesis." *Curr Opin Neurol* **14**(3): 253–8.

3. Engelbrecht, V., A. Scherer, et al. (2002). "Diffusion-weighted MR imaging in the brain in children: findings in the normal brain and in the brain with white matter diseases." *Radiology* **222**(2): 410–8.

4. Bartzokis, G., P. H. Lu, et al. (2004). "Quantifying age-related myelin breakdown with MRI: novel therapeutic targets for preventing cognitive decline and Alzheimer's disease." *J Alzheimers Dis* **6**(6 Suppl): S53–9.

5. Capello, E. and G. L. Mancardi (2004). "Marburg type and Balo's concentric sclerosis: rare and acute variants of multiple sclerosis." *Neurol Sci* **25**(Suppl 4): S361–3.

6. Levin, K. H. (2004). "Variants and mimics of Guillain Barre Syndrome." *Neurologist* **10**(2): 61–74.

7. Keirstead, H. S. (2005). "Stem cells for the treatment of myelin loss." *Trends Neurosci* **28**(12): 677–83.

8. Lewis, R. A. (2005). "Chronic inflammatory demyelinating polyneuropathy and other immune-mediated demyelinating neuropathies." *Semin Neurol* **25**(2): 217–28.

9. Totoiu, M. O. and H. S. Keirstead (2005). "Spinal cord injury is accompanied by chronic progressive demyelination." *J Comp Neurol* **486**(4): 373–83.

10. De Stefano, N., P. M. Matthews, et al. (1998). "Axonal damage correlates with disability in patients with relapsing-remitting multiple sclerosis. Results of a longitudinal magnetic resonance spectroscopy study." *Brain* **121**(Pt 8): 1469–77.

11. Bjartmar, C., X. Yin, et al. (1999). "Axonal pathology in myelin disorders." *J Neurocytol* **28**(4–5): 383–95.

12. Trapp, B. D., L. Bo, et al. (1999). "Pathogenesis of tissue injury in MS lesions." *J Neuroimmunol* **98**(1): 49–56.

13. Giulian, D. (1993). "Reactive glia as rivals in regulating neuronal survival." *Glia* **7**(1): 102–10.

14. Giulian, D., M. Corpuz, et al. (1993). "Reactive mononuclear phagocytes release neurotoxins after ischemic and traumatic injury to the central nervous system." *J Neurosci Res* **36**(6): 681–93.

15. Silver, J. (1994). "Inhibitory molecules in development and regeneration." *J Neurol* **242**(1 Suppl 1): S22–4.

16. Qiu, J., D. Cai, et al. (2000). "Glial inhibition of nerve regeneration in the mature mammalian CNS." *Glia* **29**(2): 166–74.

17. Fournier, A. E. and S. M. Strittmatter (2001). "Repulsive factors and axon regeneration in the CNS." *Curr Opin Neurobiol* **11**(1): 89–94.

18. Charles, P., R. Reynolds, et al. (2002). "Re-expression of PSA-NCAM by demyelinated axons: an inhibitor of remyelination in multiple sclerosis?" *Brain* **125**(Pt 9): 1972–9.

19. Santos-Benito, F. F. and A. Ramon-Cueto (2003). "Olfactory ensheathing glia transplantation: a therapy to promote repair in the mammalian central nervous system." *Anat Rec B New Anat* **271**(1): 77–85.

20. Smith, K. J., W. I. McDonald, et al. (1979). "Restoration of secure conduction by central demyelination." *Trans Am Neurol Assoc* **104**: 25–9.

21. Waxman, S. G., D. A. Utzschneider, et al. (1994). "Enhancement of action potential conduction following demyelination: experimental approaches to restoration of function in multiple sclerosis and spinal cord injury." *Prog Brain Res* **100**: 233–43.

22. Jeffery, N. D., A. J. Crang, et al. (1999). "Behavioral consequences of oligodendrocyte progenitor cell transplantation into experimental demyelinating lesions in the rat spinal cord." *Eur J Neurosci* **11**(5): 1508–14.

23. Totoiu, M. O., G. I. Nistor, et al. (2004). "Remyelination, axonal sparing, and locomotor recovery following transplantation of glial-committed progenitor cells into the MHV model of multiple sclerosis." *Exp Neurol* **187**(2): 254–65.

24. Keirstead, H. S., G. Nistor, et al. (2005). "Human embryonic stem cell-derived oligodendrocyte progenitor cell transplants remyelinate and restore locomotion after spinal cord injury." *J Neurosci* **25**(19): 4694–705.

25. Kornek, B., M. K. Storch, et al. (2000). "Multiple sclerosis and chronic autoimmune encephalomyelitis: a comparative quantitative study of axonal injury in active, inactive, and remyelinated lesions." *Am J Pathol* **157**(1): 267–76.

26. Zhao, C., S. P. Fancy, et al. (2005). "Stem cells, progenitors and myelin repair." *J Anat* **207**(3): 251–8.

27. Franklin, R. J., S. A. Bayley, et al. (1996). "Transplanted CG4 cells (an oligodendrocyte progenitor cell line) survive, migrate, and contribute to repair of areas of demyelination in X-irradiated and damaged spinal cord but not in normal spinal cord." *Exp Neurol* **137**(2): 263–76.

28. Franklin, R. J. and W. F. Blakemore (1997). "Transplanting oligodendrocyte progenitors into the adult CNS." *J Anat* **190**(Pt 1): 23–33.

29. Franklin, R. J., J. M. Gilson, et al. (1996). "Schwann cell-like myelination following transplantation of an olfactory bulb-ensheathing cell line into areas of demyelination in the adult CNS." *Glia* **17**(3): 217–24.

30. O'Leary, M. T. and W. F. Blakemore (1997). "Oligodendrocyte precursors survive poorly and do not migrate following transplantation into the normal adult central nervous system." *J Neurosci Res* **48**(2): 159–67.

31. Tuszynski, M. H., N. Weidner, et al. (1998). "Grafts of genetically modified Schwann cells to the spinal cord: survival, axon growth, and myelination." *Cell Transplant* **7**(2): 187–96.

32. Iwashita, Y., J. W. Fawcett, et al. (2000). "Schwann cells transplanted into normal and X-irradiated adult white matter do not migrate extensively and show poor long-term survival." *Exp Neurol* **164**(2): 292–302.

33. Ramon-Cueto, A. (2000). "Olfactory ensheathing glia transplantation into the injured spinal cord." *Prog Brain Res* **128**: 265–72.

34. Liu, S., Y. Qu, et al. (2000). "Embryonic stem cells differentiate into oligodendrocytes and myelinate in culture and after spinal cord transplantation." *Proc Natl Acad Sci U S A* **97**(11): 6126–31.

35. Ogawa, Y., K. Sawamoto, et al. (2002). "Transplantation of in vitro-expanded fetal neural progenitor cells results in neurogenesis and functional recovery after spinal cord contusion injury in adult rats." *J Neurosci Res* **69**(6): 925–33.

36. Ben-Hur, T., O. Einstein, et al. (2003). "Transplanted multipotential neural precursor cells migrate into the inflamed white matter in response to experimental autoimmune encephalomyelitis." *Glia* **41**(1): 73–80.

37. Galli, R., A. Gritti, et al. (2003). "Neural stem cells: an overview." *Circ Res* **92**(6): 598–608.

38. Iwanami, A., S. Kaneko, et al. (2005). "Transplantation of human neural stem cells for spinal cord injury in primates." *J Neurosci Res* **80**(2): 182–90.

39. Nistor, G. I., M. O. Totoiu, et al. (2005). "Human embryonic stem cells differentiate into oligodendrocytes in high purity and myelinate after spinal cord transplantation." *Glia* **49**(3): 385–96.

Chapter 6

Flow Cytometric Characterization of Neural Precursor Cells and Their Progeny

Preethi Eldi and Rodney L. Rietze

Summary

It is now clear that the adult central nervous system contains a population of neural stem and progenitor cells which act as a reservoir to underpin cell genesis for the lifetime of the animal. Unfortunately, understanding how these cells are activated both under normal conditions and following injury or disease has been a difficult task, owing not only to the rarity of these populations, but also to a paucity of cell type-specific markers. In this chapter, we will discuss in detail the methods involved in generating single cell suspension from the periventricular region of the adult mouse brain appropriate for cell sorting, and how to use negative selection strategies to produce an essentially pure population of neurosphere-forming precursor cells. While these methods have been tailored for the sorting of neural precursor cells, these methods can be easily adapted to sort for any subpopulation of neural cells based on a variety of cell surface antigen expression.

Key words: Neural stem cell, Flow Cytometry, Neurogenesis, Neurosphere, Cell Culture

1. Introduction

For more than a century, the central dogma in neuroscience touted that the brain was a static organ, incapable of cell genesis. However, it is now clear that continued and robust neurogenesis occurs in at least two discrete regions of the adult mammalian brain – the subventricular zone region (SVZ) and the hippocampal formation – putting to rest the long-standing *no new neurons after birth* dogma *(1)*. New neurons born in the SVZ migrate anteriorly along the rostral migratory stream into the olfactory bulb where they replace lost populations of granule and periglomerular interneurons, while those generated in the

Neil. J. Scolding and David Gordon (eds.), *Methods in Molecular Biology, Neural Cell Transplantation, vol. 549*
DOI: 10.1007/978-1-60327-931-4_6, © Humana Press, a part of Springer Science+Business Media, LLC 2009

subgranular zone (SGZ) of the dentate gyrus (DG) migrate into the granule cell layer and integrate into the existing neuronal network. This constant generation of new neurons originally uncovered by Joseph Altman in the 1960s *(2)* argued for the presence of a cell in the adult brain which could underpin this cell genesis for the life of the animal, namely a stem cell.

Stem cells are most generally defined as undifferentiated cells capable of extensive proliferation, self-renewal over an extended period of time, generation of large number of undifferentiated functional progeny, and regeneration of the tissue from which it was derived *(3)*. Traditionally, stem cells were thought to be located only in tissues where differentiated cells were most susceptible to loss and the need for replacement great, such as the blood, skin, and intestinal epithelia *(4)*. However, in 1992, Reynolds and Weiss employed a serum-free culture system that caused the death of the majority of cell types harvested from the periventricular region (PVR) lining the lateral ventricles of the mouse brain, but allowed a small (<0.1%) population of epidermal growth factor (EGF)-responsive stem cells to enter a period of active proliferation *(5)*. Furthermore, they were able to demonstrate that a single CNS cell could proliferate to form a ball of undifferentiated cells termed as neurosphere, which in turn could be (a) dissociated to form numerous secondary spheres or (b) induced to differentiate, generating the three major cell types of the CNS. In doing so, they showed that the cell they had isolated exhibited the cardinal stem cell attributes of proliferation, self-renewal over an extended period of time, and the ability to generate a large number of functional differentiated progeny *(3)*.

Since its original description, the neurosphere assay (NSA) has become the method of choice for the isolation, expansion, and detection of stem cell activity both in the adult and in the developing CNS. While this is of great benefit to the field, owing to the rarity of NSCs both in vivo and within the neurosphere, purification of a population of Neural stem cells (NSCs) (indeed any cell type) is necessary before one can begin to understand how the activities of such cells are regulated. To try and identify the putative NSCs, we recently harvested tissue from the PVR of the adult mouse brain, generated a single cell suspension, and sorted for distinct subpopulations of cells based on cell surface antigen expression *(6)*. While this strategy was employed by us to specifically identify NSCs, this strategy can also easily be employed to successfully purify distinct populations of more differentiated neural cell types. Accordingly, we begin here by describing the methodologies we employed to harvest SVZ tissue and generating a single cell suspension so as to optimize cell viability for the purposes of cell sorting. We conclude by describing our methodologies for the enrichment of a neural precursor cells using negative selection based on the cell surface antigens peanut agglutinin (PNA) *(7)* and heat stable antigen (HSA, also known as mCD24a) *(8)*.

2. Materials

2.1. Dissection Instruments

1. Ultra-fine curved forceps (Fine Science Tools Inc, Cat. #11251-35).
2. Small fine forceps (Fine Science Tools Inc, Cat. #11272-30).
3. Small forceps (Fine Science Tools Inc, Cat. #11050-10).
4. Small pointed scissors (Fine Science Tools Inc, Cat. #14094-11).
5. Large scissors.
6. Bead Sterilizer (Fine science Tools Inc, Cat. #18000-45).

2.2. Tissue Culture Equipment

1. Tubes:
   (a) 15-ml sterile Polystyrene Falcon tubes (Techno Plastic Products, Cat. #91015).
   (b) 50-ml sterile Polypropylene Falcon tubes (Techno Plastic Products, Cat. #91050).
   (c) Polystyrene Round-bottom test tubes 5-ml sterile (BD Falcon™ Cat. #352054).
2. Petri dishes:
   (a) 100 mm, sterile (BD Falcon™, Cat. #351029).
   (b) 35 mm, sterile (BD Falcon™, Cat. #174926).
3. 40 μm cell strainer (BD biosciences Discovery Labware, Cat. #352340).

2.3. Media Preparation

For the protocols mentioned in this chapter, commercial media as mentioned in **Subheading 2.3.1** can be used, or alternatively cost-effective media can be prepared within the laboratory as outlined in **Subheading 2.3.2**.

2.3.1. Commercial Media Preparation

Culture Reagents

1. Phosphate Buffered Saline (Stem Cell Technologies, Cat. #37350).
2. Neurocult™ NSC Basal Media (Stem Cell Technologies, Cat. #05700).
3. Neurocult™ Proliferation Supplement (Stem Cell Technologies, Cat. #05703).
4. EGF (Epidermal Growth Factor, Human recombinant, Stem Cell Technologies, Cat. #02633).
5. FGF-2 (Fibroblast Growth Factor-2, Human recombinant, Stem Cell Technologies, Cat. #02634).
6. Heparin (Sigma Aldrich, Cat. #H-3149).
7. 0.05% Trypsin–EDTA (Invitrogen-GIBCO, Cat. #25300-054).

Preparation of Hormone
Supplemented Growth
Media

To prepare 100 ml of hormone supplemented growth media; combine 90 ml of the Neurocult NSC basal media and 10 ml of the Neurocult NSC proliferation media. Store at 4°C and use within 1 week of preparation.

Preparation of Growth
Factor Stock Solutions

1. *Epidermal Growth Factor (EGF)*. Add 10 ml of the above hormone supplemented growth media to each vial of EGF to obtain stock solutions of 10 µg/ml. Store at –20°C as 50 or 100 µl aliquots.

2. *Fibroblast Growth Factor (FGF)*. Add 999 µl of hormone supplemented growth media and 1 µl BSA to each vial of bFGF to obtain stock solutions of 10 µg/ml. Store at –20°C as 50 or 100 µl aliquots.

3. *Heparin*. Dissolve 100 mg of Heparin in 50 ml of distilled water to make a stock solution of 0.2% Heparin. Filter sterilize. Store at –20°C as 1 ml aliquots.

Preparation of Complete
NSC Media

This refers to the further addition of the growth factors- EGF, FGF and Heparin to the hormone supplemented growth media. To make up 1 ml of complete NSC media, add 2 µl of the EGF stock solution, 1 µl of bFGF stock solution, and 1 µl of Heparin solution to 1 ml of hormone supplemented growth media.

*2.3.2. In-House Media
Preparation*

Outlined below is the procedure for the in-house production of cost-effective neural stem cell growth media. Ideally allocate an area strictly for media preparation to avoid contamination. Work surface should be decontaminated by 70% ethanol prior to use. Ensure that all the bottles, measuring cylinders, and other glassware used for media preparation are sterilized by autoclaving or alternatively, wherever possible, use sterile disposable labware.

Reagents Required

1. Apotransferrin. Bovine Transferrin Iron Poor (APO) (Serologicals Corporation, Cat. # 820056-1)

2. DNase-1 (Roche, Boehringer Mannheim, Cat. # 704159)

3. Dulbecco's modified Eagle's medium (DMEM), powder (4.5 g/L D-glucose) (Invitrogen-Gibco, Cat. # 12100-046)

4. EDTA (Sigma Aldrich, Cat. # E5134)

5. F12 Nutrient Mixture (Ham), Powder (Invitrogen Gibco, Cat. # 21700-075)

6. Glucose (Sigma Aldrich, Cat. # G-7021)

7. L-Glutamine, 200 mM (Invitrogen-Gibco, Cat. # 25030-024)

8. Hanks' Balanced Salt Solution with Ca2+/Mg2+ (Invitrogen, Cat. # 14025-092)

9. HEPES (Sigma Aldrich, Cat. # H-0887)

10. Insulin (Roche, Cat. # 977-420)

11. Minimum Essential medium (Invitrogen-GIBCO, Cat. # 41500-018)

12. Progesterone, 1 mg vial (Sigma Aldrich, Cat. # P-1649)

13. Putrescine (Sigma Aldrich, Cat. # P-7505)

14. Sodium Bi Carbonate (Sigma Aldrich, Cat. # S-5761)

15. Sodium Selenite (Sigma Aldrich, Cat. # S-9133)

16. Trypsin (Calbiochem, EMD Biosciences, Cat. # 6502)

17. Trypsin Inhibitor (Sigma Aldrich, Cat. # T-6522)

18. Pencillin/Streptomycin (10,000 U/ml) (Invitrogen-GIBCO, Cat. # 15140-122)

Stock Solutions to be Prepared

1. *0.1% DNase-I*. Dissolve 100 mg DNase-I in 100 ml HEM. Filter sterilize, store at –20°C as 1 ml aliquots.

2. *10× DMEM-F12*. Dissolve five 1 L packets of DMEM and F12 in 1 L of distilled water by gentle stirring. Filter sterilize and store at 4°C.

3. *30% glucose*. Dissolve 30 g glucose in 100 ml distilled water. Filter sterilize. Store at 4°C.

4. *1 M HEPES*. Dissolve 238.3 g HEPES in 1 L of distilled water. Filter sterilize. Store at 4°C.

5. *2.5 mg/ml Insulin*. Dissolve 100 mg of insulin in 4 ml 0.1N HCl. Make up the volume to 40 ml by adding 36 ml distilled water.

6. *2 mM Progesterone*. Add 1.59 ml of 95% ethanol to 1 mg vial. Mix well, store at –20°C as 100 µl aliquots.

7. *7.5% Sodium Bi-Carbonate*. Dissolve 7.52 g $NaHCO_3$ in 100 ml distilled water. Filter sterilize. Store at 4°C.

8. *3 mM Sodium Selenite*. Add 1.93 ml distilled water to 1 mg vial. Mix well, store at –20°C as 100 µl aliquots.

9. *10 mg/ml Putrescine*. Dissolve 38.6 g of Putrescine in 40 ml of distilled water.

10. *10× Hormone Mix (Alternative to Neurocult™ Proliferation Supplement)*. To 300 ml distilled water, add 40 ml 10× DMEM-F12, 8 ml of 30% glucose, 6 ml of 7.5% $NaHCO_3$, and 2.3 ml of 1 M HEPES. Mix thoroughly and add 400 mg of apotransferrin, 40 ml of 2.5 mg/ml insulin, 40 ml of 10 mg/ml putrescine stock, 40 ml of 3 mM sodium selenite and 40 µl of 2 mM progesterone. Mix the components well and store at –20°C as 10 ml aliquots.

11. *Basal Media (Alternative to Neurocult™ Basal Media)*. To prepare 450 ml of Basal Media, combine 375 ml of ultrapure distilled water, 50 ml of 10× DMEM/F12 stock solution, 10 ml of 30% Glucose, 7.5 ml of 7.5% $NaHCO_3$, 2.5 ml of

1 M HEPES, and 5 ml of 20 nM Glutamine. Mix well and filter sterilize. Store at 4°C (*see* **Note 1**).

12. *NSC Growth Medium/Hormone Supplemented Media.* To prepare 100 ml of the hormone supplemented media, combine 90 ml of the basal media with 10 ml of the 10× hormone mix. Store at 4°C.

13. *Complete NSC Medium.* To prepare 1 ml of complete NSC media, add 2 μl of the EGF stock solution, 1 μl of FGF stock solution, and 1 μl of heparin solution to 1 ml of the hormone supplemented media. (For EGF, FGF, and Heparin stock solution preparation refer **Subheading "Preparation of Growth Factor Stock Solutions"**.)

14. *Hanks Eagle Medium (HEM).* Dissolve the contents of one 10 L packet of minimum essential medium in 3 L of distilled water. In another flask add 160 ml of 1 M HEPES and 175 ml of Penicillin/streptomycin to 3 L of distilled water. Combine the two and adjust the pH to 7.2 using 10 M NaOH. Filter sterilize and store at 4°C as 100 ml aliquots. Use within 3 months of preparation.

15. *Trypsin Inhibitor.* To prepare 100 ml of trypin inhibitor, dissolve 14 mg of the trypsin inhibitor in a solution of 99 ml HEM and 1 ml of 0.1% DNase.

16. *Tissue dissociation medium (alternative to 0.05% Trypsin–EDTA).* Dissolve 40 mg EDTA, 476 mg HEPES, 1 ml 0.1% DNase-I and 50 mg of trypsin in 200 ml of Hanks' balanced salt solution (HBSS). Mix the components well and filter sterilize. Store at –20°C as 3 ml aliquots.

2.4. Antibodies

1. Peanut agglutinin – Fluorescein (PNA-FITC) (Vector Laboratories, Cat #B-1071).

2. Heat stable antigen-phycoerythrin (HSA-PE, mCD24a-PE) (BD Biosciences, Cat #553262).

3. Propidium Iodide (PI) (Sigma Aldrich, Cat #P4170): Dissolve 1 mg of Propidium Iodide in 10 ml of sterile PBS to prepare a stock solution of 100 μg/ml. Store as 500 μl aliquots at –20°C. To prepare 100 ml of PI rinse solution, reconstitute 500 μl of PI stock solution in 100 ml of NSC basal media. Filter sterilize and store at 4°C away from light.

3. Methods

Here we describe in detail the methodology involved in establishing cultures of neural precursor cells harvested from the PVR of the adult mouse brain. It has been estimated that about 1:300 cells isolated from SVZ have precursor activity based on their ability to form neurospheres. We can further enrich for the precursor sphere forming cells by cell sorting based on a combination of size and negative selection for PNA and heat stable antigen (mCD24a).

The protocol discussed below is well established with the CBA mice strain, but can be applied to various other mice strains. Prior to dissection, ensure that all the instruments, media and solutions required for the same are available. Aseptic techniques must be employed at all times to avoid contamination.

3.1. Setup

1. Sterilize all the instruments required by using either a glass bead sterilizer (250°C for 10 s) or an autoclave (120°C for 20 min). Large scissors, small pointed scissors, large forceps, and small curved forceps are required for the removal of the brain; ultrafine curved forceps and scalpel are needed for the subventricular zone dissection.

2. Prepare two small glass beakers with gauze at the bottom, filled with 70% ethanol. During dissection place the instruments in the beaker to prevent any contamination.

3. Add cold sterile Hanks Eagle Media to a 50 ml falcon tube to hold the brains after removal.

4. Place Trypsin–EDTA and culture media in a thermostatic water bath set at 37°C.

5. Sacrifice of the animals and removal of the brains must be done outside the laminar flow hood. Surface sterilize the area by wiping down with 70% ethanol and laying down several absorbent towels soaked with 70% ethanol.

6. The adult murine Subventriclar zone dissection is carried out within the laminar flow hood. Following surface sterilization, ultrafine tissue dissection instruments, three 100 mm Petri dishes and HEM are placed in the hood.

3.2. Dissection of the Adult SVZ Tissue

1. Sacrifice the mice by cervical dislocation. Rinse the head with 70% ethanol.

2. Using a pair of large scissors cuts the head just above the cervical spinal cord region.

3. Make a midline incision over the head using a small pointed scissors, reflect the skin to expose the surface of the skull. Rinse the skull with sterile HEM.

4. Make a coronal cut between the orbits of the eyes by placing each blade of the small scissors in the orbital bone after securing the head.

5. To expose the brain, make a longitudinal cut on the skull running along the sagittal suture to the Foramen magnum. Using a forceps grasp and peel the skull outward. Using the same forceps, scoop out the brain into a falcon tube containing fresh HEM.

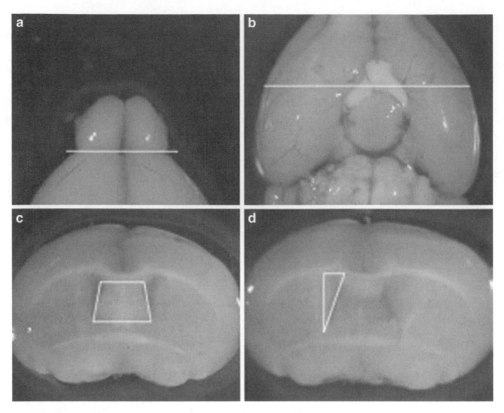

Fig. 1. Dissection of the adult murine periventricular region. (**a**) Dorsal view of the adult murine brain, illustrating the removal of the olfactory bulb (**step 7, Subheading 3.2**). (**b**) Ventral aspect of brain showing the line of coronal section along the rostral edge of the optic chaisma (**step 8, Subheading 3.2**). (**c**) The resulting section when the brain is cut along the rostral-caudal co-ordinate. The boxed area highlights the septum. (**d**) Following removal of the septum, the triangular wedge indicates the periventricular region that is harvested (**step 10, Subheading 3.2**).

6. Transfer the brains to a laminar flow hood to start the dissection of the subvenntricular zone region. Wash the brains atleast thrice with fresh HEM. Now transfer the brains into a Petri dish with little HEM ensuring that the brains are not floating for ease during dissection.

7. The first step towards the dissection of the adult SVZ tissue dissection is to remove the olfactory bulb (**Fig. 1a**). To do so, place the brain on its ventral surface and secure it at the caudal end (on either side of the cerebellum) by using a fine curved forceps. Using the second pair of the fine curved forceps, remove the olfactory bulbs.

8. Rotate the brains to expose the ventral aspect. Now using the scalpel make a coronal cut at the rostral edge of optic chaisma making sure that the cut is not oblique (**Fig. 1b**).

Transfer the cut rostral end of the brains into another Petri dish containing fresh HEM.

9. Repeat the above steps until all the brains have been sectioned.

10. Rotate the cut rostral end of the brains such that the presumptive olfactory bulb is facing downward (**Fig. 1c**). Lift out the septum using fine curved forceps and discard. Harvest a triangular wedge of the tissue along the lateral wall of the lateral ventricle making sure to exclude the striatal parenchyma and corpus callosum (**Fig. 1d**). Transfer the tissue to a new Petri dish.

11. Repeat the procedure on all sectioned brains. Remove excess HEM from the pooled dissected tissue and mince the tissue using a scalpel blade for approximately 1 min.

12. Transfer the minced tissue sample in a closed Petri dish into a tissue culture hood. Using a 1 ml filtered pipette tip and a volume of 1 ml Trypsin–EDTA per brain, transfer the tissue into a 15 ml falcon tube avoiding bubbles (*see* **Note 2**).

13. Incubate the tube for 7 min in a 37°C water bath to allow enzymatic breakdown of the tissue sample.

14. At the end of the incubation, using aseptic techniques transfer the tube to a tissue culture hood and add equal volume of trypsin inhibitor. Mix gently taking care to avoid bubbles. Centrifuge the tissue suspension at 700 rpm (102 g) for 7 min.

15. Aspirate the supernatant media. Add a suitable volume of PBS to the pellet so as to obtain a final volume of 1 ml. Using a Gilson P1000 pipette first gently mix and then triturate at-least 5–7 times until the cell suspension appears smooth. If any un-dissociated tissue bits can still be seen in the sample, then let the cell suspension settle down for 2–3 min. Carefully pipette out the top 800 µl of the cell suspension into a sterile 15 ml falcon tube avoiding the un-dissociated bits that would have settled down at the bottom. To this now add 800 µl of PBS and triturate again until the cell suspension appears smooth (*see* **Note 3**).

16. Pool the triturated samples together and make up a total volume of 10 ml using culture media. Filter this cell suspension through a 40 µm sieve and pellet the cells by centrifugation at 700 rpm (102 g) for 7 min. For further flow cytometric enrichment of the neural precursor cells, resuspend the cell pellet in 1 ml of media. Proceed to section 3.3.

17. To set up a primary culture, aspirate all the supernatant media and resuspend the cell pellet in complete NSC media to a total volume of 5 ml and seed into a T25 cm² flask. Incubate at 37°C, 5% CO_2 for 1 week (*see* **Notes 4** and **5**).

3.3. Flow Cytometric Enrichment of Neural Precursor Cells

In simple terms, during Fluorescent Activated Cell Sorting or flow cytometry, a fine stream of single cells are intercepted by laser light, the fluorescent antibodies (*see* **Note 6**) bound to the cells are identified and the population of interest sorted by application of a suitable electric field. The cells are characterized based on the size, internal complexity, and fluorescent intensity. The forward scatter (fsc) is a measure of the cell size where as the side scatter (ssc) is reflective of the internal complexity/granularity of the cells. The fsc and ssc in combination are used to exclude the debris from the sample.

Here we discuss the procedure involved in the enrichment of neural precursor cells from the adult murine SVZ region by isolating live cells >12 μm in diameter with negative selection for the PNA and the HSA antibodies as previously described *(6)*.

1. Ideally flow cytometric enrichment of the adult neural stem cells is done using pooled subventricular zone tissue harvested from 8 or 16 adult mice brains. When using 16 adult mice, process them as two separate samples of eight brains each following the above-mentioned protocol, combining both the cell suspensions in the final step in a total volume of 1 ml.

2. For the compensation controls, prepare four FACS tubes labeled cells alone, PI, PNA, and HSA containing 175 μl of complete NSC media and 25 μl of the cell suspension. Transfer the remaining cell suspension to another FACS tube labeled sort sample bringing the volume upto 1 ml using complete NSC media.

3. The antibodies mentioned in the protocol (PNA and HSA) are used at a dilution of 1:100 and 1:200 respectively. Add 2 μl of PNA–FITC to the control tube labeled PNA and 1 μl of HSA-PE to control tube labeled HSA. Add 10 μl of PNA-FITC and 5 μl of HSA-PE to the sort sample tube. Incubate all the tubes on ice in the dark for atleast 20 min for efficient antibody staining.

4. At the end of the incubation period, add 2 ml NSC media to control tubes labeled cells alone, PNA and HSA. Add 2 ml of PI rinse solution to control tube labeled PI. Add 5 ml of PI rinse solution to the sort sample tube. Mix gently using a pipette and centrifuge at 700 rpm (102 g) for 7 min.

5. Aspirate the supernatant without disturbing the cell pellet and resuspend the control tube pellets in 300 μl of complete NSC media. Resuspend the sort sample pellet in 5 ml of complete NSC media distributed over two FACS tubes. The cells are now ready to be enriched for precursor activity by flow cytometry.

6. Place the cells on ice at all times while setting up the FACS instrument. Prior to cell sorting, the fsc and ssc are adjusted to display the appropriate scatter properties as shown in **Fig. 2a**. Adjust the threshold to remove debris without affecting the

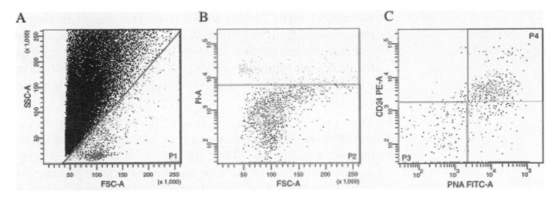

Fig. 2. (a) Representative dot plot scatter analysis of cells obtained from the adult murine subventricular zone region. Gating for cells in population 1 (P1) [1–2% of total population; tear drop population] significantly enriches for neurosphere forming cells by excluding majority of the cellular debris. (b) Representative dot plot comparing Forward scatter and Propidium Iodide staining of the P1 population. A gate is drawn around the Propidium Iodide negative population (P2) excluding the Propidium Iodide positive dead cells from further analysis. (c) Dot plot distribution of viable cells based on the HSA and PNA staining intensities. Gating for the PNA^lo HSA^lo (P3) population enriches for neural precursor cells in comparison to the PNA^hi HSA^hi population (P4).

population of interest (usually between 15,000 and 20,000 kHz). The PMT voltages on the unstained (cells alone) control tube are optimized for the required fluorescent parameters followed by compensation to correct for the spectral overlap between the different flourochromes (*see* **Notes 7** and **8**).

7. A triangular gate is drawn around the "tear drop" population as shown in **Fig. 2a** where most of the precursor activity is noted. Following exclusion of the dead PI positive cells (**Fig. 2b**), the precursor cells are greatly enriched in the PNA^lo HSA^lo population (**Fig. 2c**).

8. The cells are sorted into a 35 mm dish containing 3 ml of complete NSC media. Normally about 20,000 events of PNA^lo HSA^lo and PNA^hi HSA^hi are collected to get a valid comparable readout. Incubate at 37°C, 5% CO_2 for 7–10 days. Count the number of spheres formed when the average sphere diameter is 100–120 μm (*see* **Notes 9–12**). More sphere forming activity will be noted in the PNA^lo HSA^lo population.

4. Notes

1. The recommended shelf life of prepared Basal culture media is about 4–6 months when stored at 4°C protected from light. Hormone supplemented growth media should be used within 1 week of preparation; complete NSC media within 3 days of preparation.

2. Normally 1 ml of Trypsin-EDTA is used for enzymatic digestion of tissue from one brain. When dissecting larger number of brains for cell sorting, 3 ml of Trypsin–EDTA is sufficient for enzymatic digestion of upto 8 mice brains. Incubation with Trypsin-EDTA beyond 7 mins at 37°C will affect the cell viability.

3. Trituration refers to the process of breaking down the dissected tissue clumps to single cells by repeatedly passing them through a pipette tip. While triturating, place the pipette tip against the bottom of the tube and apply adequate pressure to restrict the flow of cells by about 50%. Care should be taken to avoid forceful trituration and generation of bubbles, as this will affect the cell viability.

4. For the optimum cell density, cells from a single brain SVZ dissection are seeded into a T-25 cm² flask, upto five brains into a T-80 cm² flask and ten brains into a T-175 cm² flask.

5. In primary cultures, all the differentiated CNS cells die in the serum free complete NSC media, only the precursor cells adhere to the substrate initially and after a few days in culture proliferate and form free floating clusters of cells termed neurospheres. Neurospheres are usually spherical in shape, have microspikes on the outer surface, appear phase bright and attain a size of 100–120 μm in 7–10 days These criteria can be used to distinguish neurospheres from aggregates of debris and dead cells which are usually found in primary cultures. The protocols mentioned in this chapter should normally generate between 300 and 600 neurospheres per mouse brain.

6. All the flourescent antibodies mentioned are light sensitive and are best stored at 4°C protected from light. Repeated Freeze-thawing will alter the antibody efficacy. Ideally aliquot antibody for use.

7. Ideally, cell sorting is done at an optimal flow rate of 5,000–7,500 cells/s with a differential pressure of 28 psi using a 90 μm nozzle and a sample pressure between 0.7 and 1 psi.

8. All the three flurochromes – FITC, PE, and PI – can be excited using the 488 laser; FITC has an emission peak at 520 nm, PE at 575 nm and PI at 617 nm.

9. To determine the cell:event ratio, in a 96-well plate containing 200 μl of complete NSC media per well collect triplicates of 200 events per population of interest. A cell count is done after allowing sufficient time for the cells to settle down (About 1–2 h).

10. To determine the cell:sphere ratio, apply the cell:event ratio to the sphere numbers obtained.

11. The sphere forming activity is greater in the PNA^{lo}HSA^{lo} population in comparison to the PNA^{hi}HSA^{hi} population, thus enriching for the precursor population. The stem cell characteristics of the population can be analyzed by studying the properties of self-renewal and multipotency.

12. The methodology described above can be used to isolate various neural cell types based on the cell surface antigen expression. For example, sorting on the basis of mCD24a can be used to purify populations of ciliated ependymal cells lining the lateral ventricle *(8)*, while F4/80 *(9)* can be used to purify microglial populations.

References

1. Gross, C.G., (2000). Neurogenesis in the adult brain: death of a dogma. *Nat Rev Neurosci*, **1**(1): p. 67–73.

2. Altman, J., (1962). Are neurons formed in the brains of adult mammals? *Science*, **135**: p. 1127–1128.

3. Potten, C.S. and Loeffler, M., (1990). Stem cells: attributes, cycles, spirals, pitfalls and uncertainties. Lessons for and from the crypt. *Development*, **110**(4): p. 1001–20.

4. Weissman, I.L., (2000). Stem cells: units of development, units of regeneration, and units in evolution. *Cell*, **100**(1): p. 157–68.

5. Reynolds, B.A. and Weiss, S., (1992). Generation of neurons and astrocytes from isolated cells of the adult mammalian central nervous system. *Science*, **255**(5052): p. 1707–10.

6. Rietze, R.L., Valcanis, H., Brooker, G.F., Thomas, T., Voss, A.K., and Bartlett, P.F., (2001). Purification of a pluripotent neural stem cell from the adult mouse brain. *Nature*, **412**(6848): p. 736–9.

7. Salner, A.L., Obbagy, J.E., and Hellman, S. (1982). Differing stem cell self-renewal of lectin-separated murine bone marrow fractions. *J Natl Cancer Inst*, **68**(4): p. 639–41.

8. Calaora, V., Chazal, G., Nielsen, P.J., Rougon, G., and Moreau, H. (1996). mCD24 expression in the developing mouse brain and in zones of secondary neurogenesis in the adult. *Neuroscience*, **73**(2): p. 581–94.

9. Perry, V.H., Hume, D.A., and Gordon, S., (1985). Immunohistochemical localization of macrophages and microglia in the adult and developing mouse brain. *Neuroscience*, **15**(2): p. 313–26.

Chapter 7

Isolation, Expansion, and Differentiation of Adult Mammalian Neural Stem and Progenitor Cells Using the Neurosphere Assay

Loic P. Deleyrolle and Brent A. Reynolds

Summary

During development and continuing into adulthood, stem cells function as a reservoir of undifferentiated cell types, whose role is to support cell genesis in several tissues and organs. In the adult, they play an essential homeostatic role by replacing differentiated cells that are lost due to physiological turnover, injury, or disease. The discovery of such cells in the adult mammalian central nervous system (CNS), an organ traditionally thought to have little or no regenerative capacity, has opened the door to the possibility of designing innovative regenerative therapeutics, an unexpected concept in neurobiology 15 years ago.

In 1992, to detect precursor cells in the adult brain, we employed a serum-free culture system whereby the majority of primary differentiated CNS cells did not survive but a small population of EGF-responsive cells were maintained in an undifferentiated state and proliferated to form clusters, called neurospheres (Reynolds and Weiss, *Science* **255**:1707–1710, 1992). These neurospheres could be (a) dissociated to form numerous secondary spheres or (b) induced to differentiate, generating the three major cell types of the CNS. This chapter outlines the adult mammalian NSC culture methodology and provides technical details of the neurosphere assay to achieve reproducible cultures.

Key words: Central nervous system, Neural stem cell, Neurospheres, Cell culture

1. Introduction

While originally debated, it is now clear that neurogenesis continues in at least two areas of the adult mammalian brain, namely, the olfactory bulb and hippocampal region *(1)*. The continuous generation of new cells strongly argues for the existence of

Neil J. Scolding and David Gordon (eds.), *Methods in Molecular Biology, Neural Cell Transplantation, vol. 549*
DOI: 10.1007/978-1-60327-931-4_7, © Humana Press, a part of Springer Science + Business Media, LLC 2009

a founder cell with the ability to proliferate, self-renew, and ultimately generate a large number of differentiated progeny, which, by definition, defines a stem cell *(2–5)*. Owing to a lack of specific and definitive markers together with poorly defined morphological characteristics, stem cells are defined based on a functional criterion so that they are, in general, defined by what they do and not by what they look like. This creates a number of practical and theoretical problems, with the most apparent being retrospective identification of a cell based on its behavior that is often elicited in an unnatural environment or conditions. This circular problem undermines many current experimental methods, and highlights the need of an alternate method of stem cell identification.

However, with this relevant caveat aside, the presence of continual cell genesis in the adult mammalian CNS necessitates the need for reasonable defining characteristics so as to establish acceptable criteria for studying and manipulating this rare and elusive cell type. Fortunately, a general well-accepted definition of a stem cell currently exists and is one that has withstood scientific scrutiny and been applied to a broad range of tissue specific stem cells. In this case, a stem cell is defined as an undifferentiated cell, which retains the capacity to (a) proliferate, (b) exhibit extensive self-renew, (c) produce a large family of differentiated functional progeny, (d) regenerate the tissue after injury, and (e) retain a flexible use of these options *(4, 6)*. Identification of a stem cell is achieved when all of these criteria are met, however, due to technical or experimental limitations, in practice only some of these criteria may be satisfied. Accordingly, characteristics such as self-renewal and multipotentiality are generally given greater credence than the other attributes when considering whether such a cell is indeed a stem cell.

In the early 1990s, we had established a culture system that allowed for the isolation and expansion, in vitro, of cells derived from the embryonic and adult CNS that exhibited key stem cell characteristics *(2, 3)*. The defined, serum-free culture system, referred to as the neurosphere assay, allowed a small population (<0.1%) of EGF-responsive cells to survive and proliferate to form a cluster of undifferentiated cells referred to as a neurosphere *(2)*. After approximately 7 days in the growth medium containing EGF, the neurospheres measure 100–200 μm in diameter and are composed of 3,000–5,000 cells (**Fig. 1a**). At this point, the spheres could be passaged and this process can be repeated every week resulting in an arithmetic increase in total numbers of cells generated. The stem cell progeny can be differentiated into the three primary CNS phenotypes – neurons, astrocytes, and oligodendrocytes, when neurospheres (either as intact clusters or dissociated cells) are plated without growth factors on an adhesive substrate (**Fig. 1b, c**). Adult-murine-derived neurospheres treated in this manner have been passaged for over 70 times,

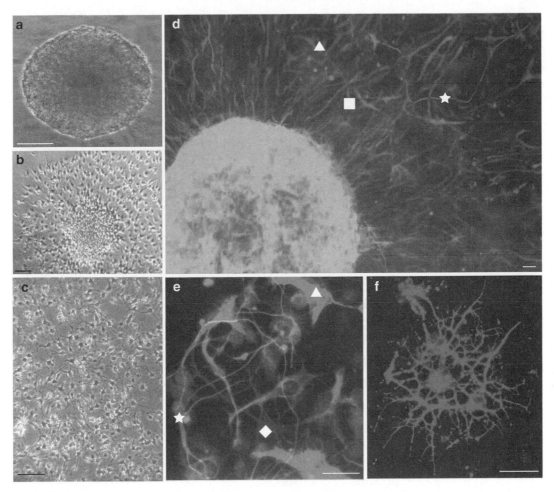

Fig. 1. EGF-responsive neural stem cells, isolated from periventricular region of adult mice were cultured in neurosphere assay. Phase contrast micrograph shows a subcultured neurosphere 6 days after passage (**a**). When transferred to differentiating conditions, neurospheres lose their spherical structure and form a monolayer of cells (**b**) expressing neuronal (β III-tubulin, ☆), astrocytic (GFAP, △) and oligodendocytic (O4, □) markers (**d**). Cells from dissociated neurospheres differentiate as well into neurons and astrocytes (☆ and △, respectively, DAPI, ◇, stains nuclei) (**e**) but also in oligodendrocytes (**f**). Scale: (**a**)–(**c**) = 50 μM; (**d**)–(**f**) = 20 μM.

resulting in a 5×10^{60}-fold increase in cell number with no loss in their proliferative or differentiation ability, growth factor responsiveness, and continued absence of tumor formation in vivo *(7)*.

Using the neurosphere assay, we, and others, have demonstrated that a population of cells exist in the fetal through to the adult mammalian CNS that can be isolated in cultures and will exhibit the critical stem cell attributes of proliferation, self-renewal, and the ability to give rise to a number of differentiated, functional progeny *(4, 8)*. While the methodology seems relatively simple to carry out, strict adherence to the procedures described here is required in order to achieve reliable and consistent results, which are essential for studying the developmental processes and elucidating the role of genetic and epigenetic factors of CNS stem cells and the determination of CNS phenotypes.

In this review, we will detail a culture methodology that allows for the isolation, propagation, and identification of stem cells from the adult mammalian brain and provide practical advice to achieve standardized, accurate, and reproducible assays.

2. Materials

2.1. General Equipment

1. Biological safety cabinet certified for level II.
2. Routine light microscope for hematocytometer cell counts.
3. Inverted microscope with flatfield.
4. Low-speed centrifuge equipped with biohazard containers.
5. 37°C incubators with humidity and gas control to maintain >95% humidity and an atmosphere of 5% CO_2 in air.
6. Pipet-aid.
7. Hematocytometer.
8. Trypan blue.

2.2. Dissection Equipment

1. Dissection microscope.
2. Large scissors.
3. Small fine scissors.
4. Small forceps.
5. Small fine forceps.
6. Ultrafine curved forceps.
7. Bead sterilizer.

2.3. Tissue Culture Equipment

1. Flasks: 25–75 cm² of 0.2-μm vented filter cap.
2. 15–50 ml polypropylene, sterile.
3. 40 μM cell strainer.

2.4. NS Media (Low Glucose Neurosphere Media)

1. *DMEM*: 1 packet (Thermo Electron Cat#50-124-PA).
2. *F12*: 1 packet (Gibco Cat#21700-026).
3. *Penicillin/Streptomycin* (Pen/Strep): 20 ml (10,000 u/ml, Gibco Cat#15140-122).
4. *NaHCO₃*: 2.25 g (Sigma Cat#S5761).
5. *Glucose*: 12 g (Sigma Cat#G7021).

6. *Hepes* 1 M: 10 ml (Sigma Cat#H4034). Note that 1 M Hepes is 238.31 g Hepes powder/1 L distilled water; filter sterilized and stored at 4°C.

7. *Distilled water*: Make up to 1.8 L.

2.5. Ingredients Used for the Growth and the Passaging of Adult Neural Stem Cells

1. *EGF*: mouse Epidermal Growth Factor, Receptor Grade (BD Biosciences Cat #354010). For a stock solution of 10 μg/ml add to one vial (100 μg) 10 ml of NS media + 10% Proliferation Supplement (Stem Cell Technologies Cat #05701). Stored in 105-μl and 205-μl aliquots at –20°C. Use 2 μl to 1 ml media (20 ng/ml final). Do *not* refreeze or use after 7 days when stored at +4°C.

2. *bFGF*: basic Fibroblast Growth Factor (Roche Cat#156Aa). For a stock solution of 10 μg/ml add to one vial (10 μg) 1 ml of NS media + 10% Proliferation Supplement (Stem Cell Technologies Cat #05701) + 2% Bovine Serum Albumin (BSA, Sigma Cat#A3311). Stored in 105-μl and 205-μl aliquots at –20°C. Use 1 μl to 1 ml media (10 ng/ml final). Do *not* refreeze or use after 7 days when stored at +4°C.

3. *Heparin*: Dissolve 50,000 units bottle (Sigma Cat#H3149) in 73.6 ml of distilled water, filtered, and stored in 1 ml frozen aliquots (stock solution: 679 u/ml). It can be kept at 4°C for 2 months or longer. Use 1 μl/ml of media (0.679 u/ml final).

4. *Trypsin/EDTA*: 0.05% (Gibco Cat#253000-054).

5. *Trypsin inhibitor solution*: 10 ml of DNase Solution (100 mg DNase (Roche Cat# 104159) dissolve in 100 ml of HEM), 0.14 g of Trypsin Inhibitor (Sigma Cat#T-6522) and HEM to make up 1 L. Use ratio 1:1 of Trypsin inhibitor solution: Trypsin/EDTA 0.05%.

6. *HEM*: MEM 1 × 10 l packet (Gibco Cat#41500-018), Pen/Strep (10,000 u/ml) 175 ml (Gibco Cat# 15140-122), 1 M Hepes 160 ml (Sigma Cat# H4034), and distilled water to make up 8.75 l.

2.6. Basal Media

1. 88-ml NS media.

2. 10 ml Proliferation Supplement (Stem Cell Technologies Cat #05701).

3. 2 ml of Bovine Serum Albumen (10% BSA, Sigma Cat#A3311).

2.7. Complete Media for the Growth of Adult Neural Stem Cells

1. 100 ml Basal Media.

2. 200 μl EGF (10 μg/ml, BD Biosciences Cat #354010).

3. 100 μl bFGF (10 μg/ml, Roche Cat#156Aa).

4. 100 μl Heparin (679 u/ml, Sigma Cat#H3149).

3. Methods

3.1. Establishment of Primary Adult Neurosphere Cultures

Here, we describe how to isolate and expand adult murine neural stem cells and to establish continuous, stem cell lines by means of growth factor stimulation.

3.1.1. Setup

Sacrifice of animals, removal and dissection of brain are performed outside the laminar flow hood. Particular caution should be exercised to avoid contamination.

1. Add cold HEM to sterile plastic Petri dishes: one or two 100-mm dishes to hold tissue; several 60-mm dishes to wash tissues; some 35-mm dishes to hold dissected tissues.

2. Warm culture medium and tissue dissociation medium to 37°C in a thermostatic waterbath.

3. Begin the dissection.

3.1.2. Dissection of Adult Periventricular Region

1. Anesthetize mice by intraperitoneal injection of pentobarbital (120 mg/kg) and sacrifice them by cervical dislocation. Tissues from two or three mice (age: from 2 to 8 months) are generally pooled to start a culture.

2. Using large scissors cut off the head just above the cervical spinal cord region. Rinse the head with 70% ethanol.

3. Using small pointed scissors make a medial caudal–rostral cut and part the skin of the head to expose the skull. Rinse the skull with sterile HEM.

4. Using the skin to hold the head in place, place each blade of small scissors in orbital bone, so as to make a coronal cut between orbits of the eyes.

5. Using the coronal cut as an entry point, make a longitudinal cut through the skull along the sagittal suture. Be careful not to damage the brain by making small cuts ensuring the angle of the blades is as shallow as possible. Cut the entire length of the skull to the foramen magnum.

6. Using curved, pointed forceps grasp and peel the skull of the each hemisphere outward to expose the brain, then using a small wetted curved spatula, scoop the brain into a Petri dish containing HEM.

7. Repeat steps 1–6 until all of the brains have been harvested.

8. Wash brains twice by subsequently transferring them to new Petri dishes containing HEM.

9. To dissect the forebrain subventricular region, place the dish containing the brain under the dissecting microscope (10× magnification). Position the brain flat on its ventral surface and hold it from the caudal side using fine curved

forceps placed on either side of the cerebellum. Use scalpel to make a coronal cut just behind the olfactory bulbs.

10. Following the removal of the olfactory bulbs, rotate the brain to expose the ventral aspect. Make a coronal cut at the level of the optic chiasm, discarding the caudal aspect of the brain.

11. Repeat steps 8–10 until all brains are sectioned.

12. Shift to a 25× magnification. Rotate the rostral aspect of the brain with the presumptive olfactory bulb facing downward. Using fine curved microscissors, first remove the septum and discard and then cut the thin layer of tissue surrounding the ventricles, excluding the striatal parenchyma and the corpus callosum. Pool dissected tissue in a newly labeled 35-mm Petri dish.

13. Upon harvesting the periventricular regions from all brains, transfer dish to tissue culture laminar flow hood. Continue to use strict sterile technique.

3.1.3. Dissociation Protocol

1. Using a scalpel blade, mince tissue for ~1 min until only very small pieces remain.

2. Add a total volume of 3 ml of tissue dissociation medium, transfer all of the minced tissues into the base of a 15-ml tube.

3. Incubate the tube for 7 min in a 37°C water bath.

4. At the end of the enzymatic incubation, return tube to hood then add an equal volume of trypsin inhibitor (3 ml).

5. Avoiding the generation of air bubbles, mix well, and then pellet the tissue suspension by centrifugation at 100 g for 5 min.

6. Vacuum off the supernatant and discard it, then resuspend the cells in 150 µl of sterile NS media containing BSA, reset the pipetter to 200 µl. Pipette up and down gently to break the clumps up.

7. Add another 800 µl of NSA so total volume is 1 ml and pass all through the 40-µm mesh filter (blue in color), which has been placed on top of a sterile 50-ml tube. (This is to remove debris.)

8. 1 brain is put into 1× T 25 flask (containing 5 ml of complete media). The cells are then incubated at 37°C, 5% CO_2 for 6–8 days by which time spheres should have formed. A count of over 300 spheres is acceptable from a T 25 flask.

3.2. Passaging Neurosphere Cultures

A variety of diameters are apparent in a bulk culture. To determine whether spheres are ready to passage, the majority of neurospheres should equal 100–150 µm in diameter (*see* **Notes 1** and **2**). If neurospheres are allowed to grow too large, they become difficult to dissociate and eventually begin to differentiate in situ.

1. If the neurospheres are ready to be passed, remove medium with suspended cells and place in an appropriate sized sterile tissue culture tube. Wash the flasks out with 2 ml warm (to prevent the cells from being shocked) basal media and add that to the centrifuge tube(s). Centrifuge cells at 100 g for 5 min at room temperature.

2. Remove the supernatant down to the actual pellet and resuspend the spheres in 1 ml of trypsin.

3. Incubate at 37°C in the water bath for 2–3 min, then inactivate the trypsin by adding an equal volume of trypsin inhibitor.

4. Mix well, but gently, to ensure that all the trypsin has been completely inactivated.

5. Centrifuge at 100 g for 4 min.

6. Vacuum off the supernatant down to the actual pellet and resuspend the cells in 1 ml of basal media and mix very well but gently.

7. Take 10 µl of this suspension and put into a 0.6-ml tube that has 90 µl of trypan blue in it. Perform a cell count. At this stage it is easy to see if the cells are single or are still aggregated, if they are not "a single cell suspension" then it will be necessary to resuspend the cells a little more vigorously. Then perform the cell count again.

8. Cells are put up at a concentration of 2.5×10^5 cells in 5 ml of complete media in a T-25 Flask.

3.3. Differentiation of Neurosphere Cultures

When cultured in the presence of EGF and/or bFGF, neural stem cells and progenitor cells proliferate to form neurospheres which, when harvested at the appropriate time-point and using the appropriate methods as described here, can be passaged practically indefinitely (7). However, upon the removal of the growth factors, neurosphere-derived cells are induced to differentiate into neurons, astrocytes, and oligodendrocytes (**Fig. 1d–f**). Overall, two methods have been described for the differentiation of neurospheres: as whole spheres cultured at low density (typically used to demonstrate individual spheres are multipotent, **Fig. 1b**) or as dissociated cells at high density (typically used to determine the relative percentage of differentiated cell types generated, **Fig. 1c**). The techniques for both methods are provided here.

3.3.1. Differentiation of Whole Neurospheres

Precoat sterile glass slides by adding a sufficient volume of poly-l-ornithine (15 mg/ml) to completely cover the glass coverslip for a period of 2 h at 37°C. Aspirate poly-l-ornithine and immediately rinse three times (10 min each) with sterile PBS (do not allow coverslips or plate to dry). Remove PBS immediately prior to the addition of neurospheres and differentiation medium.

1. Once primary or passaged neurospheres reach 150 μm in diameter, transfer the contents of the flask to an appropriate sized sterile tissue culture tube. Spin at 30 g for 5 min.

2. Aspirate essentially 100% of the growth medium, then gently resuspend (so as not to dissociate any neurospheres) with an appropriate volume of basal media.

3. Transfer neurosphere suspension to a 60-mm dish (or other sized vessel) to enable the harvesting/plucking of individual neurospheres with a disposable plastic pipette.

4. Transfer approximately ten neurospheres into individual wells of 24- or 96-well tissue culture plate containing a poly-l-ornithine coated surface with basal media.

5. After 5–8 days in vitro, individual neurospheres should have attached to the substrate and dispersed in such a manner so as to appear as a flattened monolayer of cells.

6. Proceed to fix cells with the addition of 4% paraformaldehyde (in PBS, pH 7.2) for 20 min at room temperature and then process the adherent cells for immunocytochemistry as required.

3.3.2. Differentiation of Dissociated Cells

1. Once primary or passaged neurospheres reach 150 μm, transfer the contents of the flask to an appropriate sized sterile tissue culture tube. Spin at 30 g for 5 min.

2. Remove essentially 100% of the supernatant and resuspend cells using 1 ml of trypsin/EDTA, incubating at room temperature for 3–4 min.

3. Add 1 ml of trypsin inhibitor to each tube, mix well, and then centrifuge cell suspension(s) at 100 g for 5 min.

4. Remove essentially 100% of the supernatant and resuspend cells by the addition of 1 ml of basal media. Triturate cells until suspension appears milky and no spheres can be seen (~5–7 times).

5. Combine a 10-μl aliquot from the cell suspension with 90 μl of Trypan blue in a microcentrifuge tube, mix, and then transfer 10 μl to a hemocytometer so as to perform a cell count.

6. Seed individual wells of 24-well tissue culture plate containing a poly-l-ornithine coated glass coverslip with 5×10^5 cells.

7. After 4–6 days in vitro, neurosphere-derived cells will have differentiated sufficiently. Proceed to fix cells with the addition of 4% paraformaldehyde (in PBS, pH 7.2) for 20 min at room temperature and then process the adherent cells for immunocytochemistry as required.

3.4. Immunostaining of Differentiated Cells

3.4.1. Labeling with Primary Antibodies

1. Remove the paraformaldehyde solution using an aspiration system connecting to a vacuum pump.
2. Add PBS (pH 7.2) to the samples and incubate for 5 min. Aspirate PBS using vacuum pump and repeat this washing procedure two more times for a total of three wash steps.
3. Permeabilize and block cells for 60 min in PBS-0.1% Triton-X100 + 5% Normal Goat Serum at 37°C.
4. Incubate cells for 60–90 min at room temperature with primary antibodies diluted in blocking solution (or over night at 4°C) (*see* **Table 1**).
5. Wash cells 3× with PBS.

3.4.2. Secondary Antibody Staining

1. Prepare a 1:400 dilution of the secondary antibodies diluted in blocking buffer.
2. Incubate 30–60 min at 37°C.
3. Wash cells three times with PBS, include DAPI (1:1000) in second wash for nuclear counterstain.

Table 1
Suggested primary antibodies and targeted antigens for the different neural lineages

Targeted antigen	Primary antibody		
	Isotype	Working dilution	Catalog np.
Neurons			
Neuronal class B-tubulin	Mouse IgG	1:1,000	Promega #G7121
Microtubule-associated protein-2 (MAP2)	Mouse IgG	1:300	Chemocon #MAB3418
Double cortin	Guinea pig polyclonal	1:1,000	Chemicon #AB5910
Astrocytes			
Glial fibrillary acidic protein (GFAP)	Rabbit polyclonal	1:500	DakoCytomation # Z0334
Oligodendrocytes			
Oligodendrocytes marker	Mouse IgM	1:50	Chemicon #MAB345
Myelin basic protein (MBP)	Rabbit Polyclonal	1:200	Chemicon #AB980

4. Mount on slides using DAKO fluorescent mounting media (S3023).

5. Visualize immunostaining under fluorescent microscope using appropriate filters for fluorophore.

6. **Figure 1d** shows the differentiation of an undissociated neurosphere in neurons (β III-tubulin staining, ☆), astrocytes (GFAP, △), and oligodendrocytes (O4, □) whereas **Fig. 1e, f** shows neuronal, astrocytic, and oligodendrocytic labeling in differentiated-dissociated neurosphere culture.

4. Notes

1. Spheres must be rounded but not compacted; they should measure between 100 and 150 μm (**Subheading 3**; **Fig. 1a**).

2. Primary neurospheres are often associated with cellular debris; however, subculturing will effectively select for proliferating precursor cells and remove cell aggregates, debris, and dead cells (**Subheading 3**).

References

1. Gross, C. G. (2000). Neurogenesis in the adult brain: death of a dogma. *Nat Rev Neurosci* **1**, 67–73.

2. Reynolds, B. A., and Weiss, S. (1992). Generation of neurons and astrocytes from isolated cells of the adult mammalian central nervous system. *Science* **255**, 1707–10.

3. Reynolds, B. A., and Weiss, S. (1996). Clonal and population analyses demonstrate that an EGF- responsive mammalian embryonic CNS precursor is a stem cell. *Dev Biol* **175**, 1–13.

4. Potten, C. S., and Loeffler, M. (1990). Stem cells: attributes, cycles, spirals, pitfalls and uncertainties. Lessons for and from the crypt. *Development* **110**, 1001–20.

5. Reynolds, B. A., and Rietze, R. L. (2005). Neural stem cells and neurospheres-re-evaluating the relationship. *Nat Methods* **2**, 333–6.

6. Evans, G. S., and Potten, C. S. (1991) Stem cells and the elixir of life. *Bioessays* **13**, 135–8.

7. Foroni, C., Galli, R., Cipelletti, B., Caumo, A., Alberti, S., Fiocco, R., and Vescovi, A. (2007). Resilience to transformation and inherent genetic and functional stability of adult neural stem cells ex vivo. *Cancer Res* **67**, 3725–33.

8. Hall, P. A., and Watt, F. M. (1989). Stem cells: the generation and maintenance of cellular diversity. *Development* **106**, 619–33.

Chapter 8

Human Mesenchymal Stem Cell Culture for Neural Transplantation

David Gordon and Neil J. Scolding

Summary

Mesenchymal stem cells (MSCs) have the potential to play a role in autologous repair of central nervous system injury or disease, circumventing both the complications associated with immune rejection of allogenic cells, and many of the ethical concerns associated with embryonic stem cell use. Human bone marrow-derived MSCs can be extracted relatively simply from the marrow of adult patients and maintained and expanded in culture. More importantly, it has been previously demonstrated that MSCs have the capacity to differentiate into neurons and glia *in vitro* when grown under appropriate conditions. Multipotent MSCs have also been successfully used in transplantation studies in animal models of disease as diverse as demyelination, stroke, trauma and Parkinson's disease. MSCs therefore provide an attractive and practical source of stem cells for reparative therapy in patients, and in this paper we describe methods for the reproducible culture and neural differentiation of human MSCs generated from patient marrow.

Key words: Culture, Neural differentiation, Human mesenchymal stem cells

1. Introduction

Stem cells have enormous potential as therapeutic tools in the treatment of neurological diseases as diverse as multiple sclerosis (MS), stroke, trauma and Parkinson's disease (*1–4*). Most studies have focused on the use of embryonic tissue as the primary stem cell source, but this raises significant practical, immunological and ethical concerns. It is now well accepted that adult stem cells are present in a variety of different tissue types and that these tissue-specific cells have the capacity to differentiate into a wider range of cell types than previously thought (*5–8*). The multipotentiality of adult stem cells has generated much interest in their use as autologous treatments for neurological disease, since it provides

Neil J. Scolding and David Gordon (eds.), *Methods in Molecular Biology, Neural Cell Transplantation, vol. 549*
DOI: 10.1007/978-1-60327-931-4_8, © Humana Press, a part of Springer Science+Business Media, LLC 2009

a way to circumvent many of the difficulties arising from the use of embryonic stem cells (*5, 6, 9, 10*).

An attractive source of adult stem cells for autologous stem cell replacement therapies in the central nervous system (CNS) is bone marrow (*11–14*). Human bone marrow-derived mesenchymal stem cells (MSCs) can be extracted relatively simply from adult tissue and maintained in cell culture. When cultured under the appropriate conditions MSCs are able to generate cells specific to the mesenchymal lineage *in vitro*, specifically chondrocytes (cartilage), osteoblasts (bone) and adipocytes (fat) (*12, 14*). It has further been demonstrated that both human and rodent MSCs display remarkable phenotypic plasticity, *trans*-differentiating into neural cells *in vitro* in the presence of appropriate growth factors (*15–18*). More recently it has been shown that human MSCs cultured in the presence of epidermal growth factor (EGF) and basic fibroblast growth factor (bFGF) can be induced to grow as neural stem cell (NSC)-like neurospheres *in vitro* (*19*). The cells within these neurosphere-like structures express high levels of early neuroectodermal markers, such as the proneural genes and nestin (*19*).

Such studies have provided the impetus for a range of experimental investigations exploring the potential reparative or neuroprotective effects of MSCs in models of neurological or neurodegenerative disease (*20–23*). MSCs therefore represent an attractive clinical option for developing cell therapies for disease treatment. However, if such studies successfully lead to future translational studies in patients, then human rather than rodent cells must be investigated, and in this paper we describe practical methods for the culture and differentiation of such human MSCs.

2. Materials

2.1. General Lab and Cell Culture Equipment

1. Class 2 biological safety cabinet or laminar-flow hood.
2. Vacuum pump and trap containing disinfectant.
3. Inverted light microscope with phase contrast optics for checking cells and cell counts.
4. Falcon-style 15-ml and 50-ml tubes.
5. Plastic Petri-dishes and disposable scalpels.
6. Sterile disposable glass Pasteur pipettes.
7. Sterile disposable plastic Pasteur pipettes.
8. Haemocytometer and Trypan blue solution (Sigma-Aldrich) for cell counts.

9. Cell culture incubators set at 37°C and 5% CO_2 in air, and able to maintain >95% humidity.

10. Bench top, low speed centrifuge with an option to enable slow deceleration. Buckets that can accommodate 15-ml and 50-ml tubes.

11. Water-bath set at 37°C.

12. P2, P20, P200, P1000 Gilson Pipetman pipettes or equivalent.

2.2. Human MSC Culture

1. RPMI-1640 + 10% heparin. Add 5 ml of heparin solution to 45 ml of RPMI-1640 to give a final concentration of 100 U heparin/ml. Ten milliliters are sufficient in a sterile 50-ml tube for marrow collection.

2. Hanks' balanced salt solution (HBSS).

3. Ficoll.

4. Red blood cell lysis buffer. Add 4.15 g ammonium chloride (NH_4CL), 0.5 g potassium bicarbonate ($KHCO_3$) and 0.02 g EDTA to 500 ml of distilled H_2O. Mix and filter sterilise. Can be frozen for long-term storage.

5. Dulbecco's modified Eagles medium (DMEM). Add 10 g of DMEM powder and 3.7 g of sodium bicarbonate ($NaHCO_3$) to 1 L of distilled H_2O. Mix, pH to 7.2 and filter sterilise.

6. Phosphate buffered saline (PBS), pH 7.4. Autoclave or filter sterilise.

7. Foetal bovine serum (FBS).

8. Recombinant human basic fibroblast growth factor (rh bFGF). Store as 25 μg/ml stocks made up in PBS + 1% FBS.

9. 25-cm^2, 75-cm^2 and 175-cm^2 tissue culture flasks.

10. Trypsin/EDTA solution.

11. Accutase (Chemicon).

12. FBS for human MSCs (Stem cell technologies).

13. MSC growth medium. 50 ml of FBS for human MSCs, 10 ml of penicillin/streptomycin solution, 2.5 ml of a 200 mM solution of L-glutamine and 10 μl of a 25 μg/ml stock of rh bFGF (1 ng/ml final concentration) in 437.5 ml of DMEM. Filter sterilise.

14. 250-ml (Corning Incorporated) and 500-ml (Corning Incorporated) filter systems.

15. 13-mm diameter round coverslips.

16. 4-well and 24-well tissue culture plates.

17. Poly-L-lysine. Mix with 50 ml of distilled H_2O. Filter sterilise.

18. Paraformaldehyde. Dilute to a 4% solution in PBS.

2.3. Long-Term Storage of Human MSC Cultures

1. Cryotubes.
2. MSC growth medium + 50% FBS. Add 25 ml of MSC growth medium and 25 ml of FBS. Filter sterilise.
3. MSC growth medium + 20% DMSO. Add 40 ml of MSC growth medium and 10 ml of dimethyl sulphoxide (DMSO).
4. Isopropanol.
5. Cryo 1°C freezing container (Nalgene) containing isopropanol.

2.4. Confirming the Mesenchymal Origin of Human MSCs

1. Propidium Iodide solution.
2. FACS tubes.
3. Adipogenic induction medium. Add 5 ml of FBS, 1 ml of penicillin/streptomycin solution, 25 µl of a 1 mM stock of 1-methyl-3-isobutylxanthine (final concentration of 0.5 µM), 50 µl of a 1 mM stock of dexamethasone (final concentration of 1 µM), 50 µl of a 10 mg/ml stock of insulin (final concentration of 10 µg/ml) and 0.179 g of indomethacin (final concentration of 100 µM) to 44 ml of DMEM. Filter sterilise.
4. Adipogenic maintenance medium. Add 5 ml of FBS, 1 ml of penicillin/streptomycin solution and 50 µl of a 10 mg/ml stock of insulin (final concentration of 10 µg/ml) to 44 ml of DMEM. Filter sterilise.
5. Osteogenic induction medium. Add 5 ml of FBS, 1 ml of penicillin/streptomycin solution, 250 µl of a 10 mM stock of ascorbic acid (final concentration of 0.05 mM), 5 µl of a 1 mM stock of dexamethasone (final concentration of 100 nM) and 0.108 g of glycerol 2-phosphate (final concentration of 10 mM) to 44 ml of DMEM. Filter sterilise.
6. Chondrogenic induction medium. Add 12.5 µl of a 2 µg/ml stock of recombinant human transforming growth factor-β3 (final concentration of 10 ng/l), 1 ml of penicillin/streptomycin solution, 3.5 µl of a 1 mM stock of dexamethasone (final concentration of 10^{-7} M), 500 µl of a 10 mM stock of ascorbic acid (final concentration of 0.1 mM), 125 µl of a 10 mg/ml stock of insulin (final concentration of 25 µg/ml) and 0.0625 g of bovine serum albumin (BSA) (final concentration of 1.25 mg/ml) to 49 ml of DMEM. Filter sterilise.
7. 0.5 M sucrose solution. Add 17.115 g of sucrose to 100 ml of PBS. Mix well.
8. OCT.

2.5. The Neurogenic Capacity of Human MSCs

1. Low attachment plates and flasks.
2. Recombinant human epidermal growth factor (rh EGF). Store as 2 mg/ml stocks made up in PBS + 0.1% BSA.
3. Neurosphere medium. Add 5 ml of N2 supplement (final concentration of 1%; Invitrogen), 10 ml of penicillin/streptomycin

solution (Gibco), 5 ml of a 200 mM solution of L-glutamine (final concentration of 2 mM), 1 ml of a 10 mg/ml stock of insulin (final concentration of 25 µg/ml), 5 µl of a 2 mg/ml stock of rh EGF (final concentration of 20 ng/ml; Sigma-Aldrich), 400 µl of a 25 µg/ml of rh bFGF (final concentration of 20 ng/ml; Sigma-Aldrich) and 100 µl of a 10 µg/ml stock of human leukaemia inhibitory factor (final concentration of 2 ng/ml; Sigma-Aldrich) to 480 ml of Neuromed-N (Euroclone, UK). Filter sterilise.

4. Neural induction medium. Add 500 µl of FBS (final concentration of 1%), 2.5 ml of horse serum (final concentration of 1%), 10 µl of a 25 mM stock solution of retinoic acid (final concentration of 5 µM) and 500 µl of N2 supplement (final concentration of 1%) to 46.5 ml of Neurobasal medium (Gibco). To promote predominantly neuronal differentiation, add 50 µl of a 10 µg/ml stock of recombinant human brain derived neurotrophic factor (rh BDNF) (final concentration of 10 ng/ml; Peprotech, UK). To promote predominantly glial differentiation, add 50 µl of a 10 µg/ml stock of recombinant human platelet derived growth factor BB (rh PDGF-BB) (final concentration of 10 ng/ml; R and D systems). Filter sterilise.

2.6. Immunostaining

1. Blocking solution. Add 2 ml of normal goat serum (final concentration of 4%) and 50 µl of Triton-X100 (final concentration of 0.1%) to 48 ml of PBS.

2. 22 × 22 mm coverslip.

3. Vectashield containing DAPI.

3. Methods

3.1. Human MSC Culture

3.1.1. Extraction of Human MSCs from Bone Marrow

1. Human bone marrow is collected from the femoral neck of patients undergoing total or partial hip replacement. Marrow is placed into a sterile 50-ml tube containing RPMI-1640 containing 10% heparin solution to prevent clotting (*see* **Note 1**).

2. The samples are left overnight at room temperature in the dark.

3. Transfer the marrow sample from the collection tube to a Petri dish. Add a small volume of HBSS (approximately 5 ml) to prevent the sample drying out. Using sterile scalpels break up the marrow as finely as possible, and scrape marrow from the underlying bone.

4. When satisfied that as much of the sample as possible has been collected, transfer the cell-containing HBSS to a fresh, sterile

50-ml tube. At this stage, the homogenised cell suspension will appear as thick slurry, and may contain small particles of bone and debris. It is easiest to transfer the solution with a 10-ml disposable serological pipette. Rinse the Petri dish 2–3 times with HBSS to collect cells, and transfer this also to the sterile 50-ml tube.

5. Bring the total volume of the cell suspension up to 50 ml with HBSS.

6. For every 25 ml of diluted cell suspension, layer this over 25 ml of Ficoll in fresh 50-ml tubes. To prevent mixing at the solution interface, the tube should be held at a 45° angle and cells added by gently running the suspension along the inside wall of the tube using a disposable plastic Pasteur pipette.

7. Samples are then centrifuged at $1,200 \times g$ for 30 min. It is important that slow deceleration is activated on the centrifuge to prevent mixing at the solution interface while the centrifuge is coming to a halt.

8. Following centrifugation, the mononuclear cell layer is aspirated from below the marrow/Ficoll interface and transferred to a fresh 50-ml tube (*see* **Note 3**). The mononuclear cell layer will appear white and cloudy, and is most easily removed by passing a disposable plastic Pasteur pipette gently through the solution until the tip is just below the interface, and drawing up the mononuclear cell layer.

9. Add 30 ml of HBSS to the aspirate and centrifuge at $300 \times g$ for 10 min. Ficoll is toxic to cells, so it important to wash cells sufficiently to remove excess Ficoll.

10. Resuspend the resulting pellet again in 30 ml of HBSS, and centrifuge at $300 \times g$ for 10 min.

11. Resuspend the cell pellet in red blood cell lysis buffer at 4°C for 10 min, followed by centrifugation at $300 \times g$ for 10 min.

12. Wash the resultant pellet with 20 ml HBSS, and centrifuge at $300 \times g$ for 10 min.

13. Repeat **step 11**.

14. Resuspend the final pellet in 1–5 ml of MSC growth medium (*see* **Note 2**).

15. Count the number of cells using a haemocytometer. Plate the cells at 1×10^7 cells per 25-cm² tissue culture flask in MSC growth media (*see* **Note 4**).

16. After 48 h replace media with fresh MSC growth medium to remove nonadherent cells.

17. Media should then be replaced with fresh MSC growth medium every 3 days.

18. Cells should be confluent and ready for passage after 7–10 days (**Fig. 1a**).

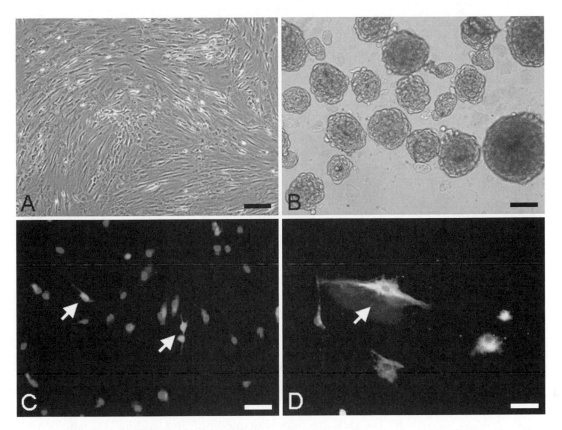

Fig. 1. Human MSCs under different culture conditions. Human MSCs grown to confluence in MSC growth medium display a flattened, fibroblast appearance under phase contrast (**a**). When cultured in neurosphere media containing EGF and bFGF, MSCs generate NSC-like neurospheres of different diameters (**b**). When cultured under conditions promoting neural differentiation, human MSCs express neural markers. Staining for βIII-tubulin identifies cells with a predominantly bipolar appearance and the presence of process-like extensions from the poles of the cell body (**c**; *arrows*). GFAP-labeled cells (*arrow*) exhibit intense staining throughout the cytoplasm and a predominantly flattened morphology (**d**). Scale bars 40 μm.

3.1.2. MSC Passage

1. Remove old MSC growth medium from flasks by aspiration, and rinse the cells with 2 × 10 ml of PBS to remove serum and cell debris.

2. To dissociate adherent cells from the surface of the TC flask, either trypsin/EDTA or Accutase can be used. Depending on the size of the tissue culture flask, add the appropriate volume of trypsin/EDTA or Accutase solution. This is approximately 1 ml for a 25-cm² flask, 3 ml for a 75-cm² flask and 7 ml for a 175-cm² flask.

3. Incubate the flasks at 37°C for 8 min.

4. To ensure that all adherent cells have detached from the surface, the flask can be sharply tapped on a table surface several times to dislodge cells. Care must be taken not to splash media on the upper surface of the flask when doing this.

5. The resulting cell suspension is transferred to a sterile 15-ml tube. If trypsin/EDTA is used to dissociate cells, add approximately

3× volume of a 15% solution of FBS in HBSS to inhibit trypsin. Accutase is self-digesting and therefore does not require the addition of an inhibitor solution.

6. Centrifuge the suspension at $300 \times g$ for 8 min in a bench top centrifuge.

7. Resuspend the final pellet in 1 ml of MSC growth medium.

8. Count the number of MSCs using a haemocytometer and plate them at 1×10^5 cells per 25-cm^2 flask, 5×10^5 cells per 75-cm^2 flask or 1.2×10^6 cells per 175-cm^2 flask in MSC growth medium.

3.1.3. Growth of MSCs on Coverslips

1. Sterile 30-mm diameter round coverslips are placed into each well of a suitably sized tissue culture plate (4- or 24-well plate).

2. Add 500 µl of poly-L-lysine solution and incubate at room temperature for a minimum of 30 min.

3. Excess poly-L-lysine is removed and the coverslips dried in a 60°C drying oven overnight.

4. Passage MSCs as described until the final cell suspension is generated. Plate MSCs at approximately 1×10^4 cells per well in MSC growth media. Cells will be confluent 5–7 days after initial setup.

5. Fix cells by removing medium and adding 4% paraformaldehyde for 30 min at 37°C. Remove paraformaldehyde and rinse coverslips with several washed of PBS. Coverslips can be stored in PBS for several weeks prior to immunostaining if required.

3.2. Long-Term Storage of Human MSC Cultures

3.2.1. Cryopreserving MSCs

1. Passage MSCs as described to generate a single cell suspension and calculate the number of cells using a haemocytometer.

2. Transfer a minimum of 1×10^5 cells to each cryotube

3. To each cryotube, slowly add 500 µl of prewarmed 50% MSC growth medium + 50% FBS. Gently mix the suspension.

4. To each cryotube, add a further 500 µl of prewarmed MSC growth medium + 20% DMSO. Very gently mix the suspension to prevent any spillage.

5. Seal the cryotubes and transfer to a controlled-rate freezing container. The use of isopropanol in these containers results in a controlled freezing rate of 1°C/min, and prevents snap freezing which can damage cells (*see* **Notes 5** and **6**).

6. Place the controlled-rate freezing container into a –80°C freezer, and leave overnight.

7. Cryotubes can then be transferred to liquid nitrogen for long-term storage.

3.2.2. Thawing Cryopreserved Cells

1. Cryotubes are removed from liquid nitrogen and thawed in a waterbath at 37°C at least 10 min prior to use.

2. Transfer 10 ml of prewarmed MSC growth medium to sterile 15-ml tubes. Label each tube to match a corresponding cryotube.

3. When the cryotubes are thawed, transfer them to the sterile biocabinet.

4. Using a 10-ml serological pipette, take up 3–4 ml of MSC growth medium from a 15-ml tube. Slowly mix the MSC growth medium and the thawed cell suspension, ensuring that all cells are rinsed from the tube. Care must be taken not to spill any medium, as this can lead to contamination.

5. Transfer the 3–4 ml of medium, which now contains the thawed MSCs, into the medium that remains in the 15-ml tube.

6. When all samples have been transferred from cryotubes into the appropriately-labeled 15-ml tubes, the samples are centrifuged at $300 \times g$ for 8 min in a bench top centrifuge.

7. Aspirate the supernatant from each 15-ml tube, and resuspend each pellet in 10 ml of fresh MSC growth medium.

8. Again centrifuge the sample at $300 \times g$ for 8 min.

9. After the second spin, each pellet is resuspended in 5 ml of MSC growth medium and transferred to a 25-cm² TC flask. Cells are then incubated and passaged as normal.

3.3. Confirming the Mesenchymal Origin of Human MSCs

An important consideration when working with MSCs is confirmation that such cells are indeed of mesenchymal origin. MSCs can be identified by flow cytometry by expression of a repertoire of surface receptors (either positive or negative surface expression depending on the marker) (*14*). Differentiating MSCs under conditions that induce cells to follow a specific developmental pathway (*14*) can also confirm their multilineage capacity. Following treatment, the ability of cells to appropriately differentiate can be examined by histochemical or immunocytochemical staining to detect the presence of suitable markers (*12, 14*).

3.3.1. Flow Cytometry of Surface CD Markers

1. Passage MSCs as described to generate a single cell suspension and calculate the number of cells using a haemocytometer.

2. Add 5% normal goat serum to the suspension to block cells. Incubate for 30 min at room temperature.

3. Transfer a minimum of 5×10^4 cells to each sterile 15-ml centrifuge tube required. Add sufficient PBS to bring the volume to 1 ml.

4. Label the tubes as antibody, isotype control, secondary antibody alone, and cells alone.

5. Add the anti-CD marker primary antibody to the appropriately labeled tube and incubate for 1 h on ice in the dark (**Table 1**).
6. Add 10 ml of PBS to each tube, and centrifuge at 300 × g for 8 min to rinse cells.

Table 1
Some common primary antibodies for human mesenchymal stem cell study

| Markers | Antibody details | | |
	Isotype	Dilution	Supplier (cat. no.)
MSC-expressed			
Fibronectin	Mouse IgM	1:400	Sigma-Aldrich (F6140)
Vimentin	Mouse IgM	1:200	Sigma-Aldrich (V5255)
CD (surface)			
CD44	Mouse IgG$_1$	1:100	AbD Serotec (MCA89XZ)
CD45	Mouse IgG$_1$	1:100	AbD Serotec (MCAP87)
CD90	Mouse IgG$_1$	1:100	AbD Serotec (MCAP90T)
CD105	Mouse IgG$_1$	1:100	AbD Serotec (MCA1557)
CD166	Mouse IgG$_1$	1:100	AbD Serotec (MCA1926GA)
Adipogenic			
Lipoprotein lipase	Mouse IgG$_1$	1:200	Abcam (Ab21356)
Osteogenic			
Alkaline phosphatase	Mouse IgG$_1$	1:200	Abcam (Ab17973)
Chondrogenic			
Human Collagen II	Mouse IgG$_1$	1:200	Chemicon (MAb1330)
Neurogenic			
Nestin	Mouse IgG$_1$	1:500	Chemicon (MAb5326)
βIII-tubulin	Rabbit polyclonal	1:1000	Sigma-Aldrich (T2200)
Glial fibrillary acidic protein (GFAP)	Mouse IgG$_1$	1:500	Chemicon (MAb360)
Oligodendrocyte marker O4	Mouse IgM	1:100	Chemicon (MAb345)
Myelin basic protein (MBP)	Rat IgG$_{2a}$	1:10	AbD Serotec (MCA409S)

7. Resuspend the pellets in 1 ml of PBS. Incubate the cells with secondary antibody or isotype control. The sample labeled cells alone gets neither. Incubate the cells for 1 h on ice in the dark.

8. Add 10 ml of PBS to each tube, and centrifuge at $300 \times g$ for 8 min rinse cells.

9. Resuspend the cells in 1 ml of propidium iodide solution and incubate for 5 min on ice in the dark.

10. Add 10 ml of PBS to each tube, and centrifuge at $300 \times g$ for 8 min rinse cells.

11. Resuspend the pellets in 1 ml of PBS and transfer to FACS tubes. Store the samples on ice until ready to run through the flow cytometer. The cells alone, secondary antibody alone, and isotype control sample are all used to set initial flow cytometry gates, while propidium iodide is used to exclude all dead cells and debris.

3.3.2. Adipogenic Induction

1. Cells are grown to confluence on poly-L-lysine coated coverslips in MSC growth medium over 7–10 days.

2. Medium is removed from the wells, and cells rinsed with PBS × 2 to remove excess growth media.

3. Add 500 μl of adipogenic *induction* medium to each well.

4. Incubate the cell culture plates for 3 days.

5. Aspirate off the media, and add 500 μl of adipogenic *maintenance* medium to each well.

6. Addition of adipogenic induction and maintenance media is alternated every 3 days until the presence of lipid vacuoles are observed under the phase contrast microscope. It should take 14–21 days for this to occur.

7. Fix cells by removing medium and adding 4% paraformaldehyde for 30 min at 37°C. Remove paraformaldehyde and rinse coverslips with several washed of PBS. Coverslips can be stored in PBS for several weeks prior to histochemical staining for lipids or immunostaining (**Table 1**).

3.3.3. Osteogenic Induction

1. Cells are grown to confluence on poly-L-lysine coated coverslips in MSC growth medium over 7–10 days. All incubations are carried out at 37°C/5% CO_2.

2. Remove the old medium from the wells and rinse the cells and with PBS × 2 to remove excess growth media.

3. Add 500 μl of osteogenic induction medium to each well.

4. Replace with fresh osteogenic induction medium every 3 days for 3–4 weeks.

5. Fix cells by removing medium and adding 4% paraformaldehyde for 30 min at 37°C. Remove paraformaldehyde and

rinse coverslips with several washed of PBS. Coverslips can be stored in PBS for several weeks prior to histochemical staining for bone markers or immunostaining **(Table 1)**.

3.3.4. Chondrogenic Induction

1. Following MSC passage and generation of a cell suspension, transfer 1.5×10^5 cells to a sterile 1.5-ml eppendorf tube.

2. Centrifuge the cells at $300 \times g$ for 8 min.

3. Remove the supernatant, and add 500 µl of chondrogenic induction medium

4. Replace with fresh chondrogenic induction medium every 3 days for 3–4 weeks. The cells will form a pelletted micromass, so care must be taken when removing old medium that the pellet remains at the bottom of the tube.

5. When ready, fix the pellet for 30 min in 4% paraformaldehyde.

6. After several washes with PBS to remove excess fixative, cryo-protect the sample by submersion overnight in 0.5 M sucrose at 4°C.

7. Embed pellets in OCT compound in plastic moulds, and allow the blocks to set at –20°C.

8. Cut 12–20 µm thick sections in a cryostat containing a microtome. Mount the sections onto microscope slides and examine the sections for the presence of chondrogenic markers using histochemical staining for cartilage proteoglycans or immunostaining **(Table 1)**.

3.4. The Neurogenic Capacity of Cultured MSCs

3.4.1. Generation of NSC-Like Neurospheres from Human MSCs

1. MSCs are harvested as normal from culture flasks during cell passage.

2. Cells are plated down in low-attachment 12-well plates at 1×10^4 cells/well, or into low-attachment 75-cm² flasks at 5×10^5 cells/flask in neurosphere medium.

3. Cultures are incubated under normal conditions for 7–10 days. When cultures are ready, NSC-like neurospheres should range in size from 100 to 120 µm **(Fig. 1b)**.

4. When ready for passage, collect the NSC-like neurospheres by transferring the supernatant to a sterile 15-ml tube. Add 2 ml of prewarmed PBS or neurosphere medium to the flask to rinse out any remaining neurospheres, and add this to the medium already collected.

5. Centrifuge the samples at $100 \times g$ for 8 min in a bench top centrifuge. Discard the supernatant, and resuspend the pellet in 1 ml of either trypsin/EDTA or Accutase.

6. Incubate the samples in a 37°C waterbath for 5 min.

7. If trypsin/EDTA is used to dissociate cells, add approximately 3× volume of a 15% solution of FBS in HBSS to inhibit trypsin,

and mix well. Accutase is self-digesting and therefore does not require the addition of an inhibitor solution.

8. Centrifuge samples at $100 \times g$ for 8 min, and discard the supernatant.

9. Resuspend the pellet in 1-ml of neurosphere medium, and gently triturated 5× through the tip of a Gilson P200 to generate a single-cell suspension.

10. Count the number of cells using a haemocytometer, and replate as described in **step 2**.

3.4.2. Fixation of Intact NSC-Like Neurospheres

It can be advantageous to immunostain intact NSC-like neurospheres, to examine the distribution of neural markers within the neurosphere.

1. Collect NSC-like neurospheres by transferring the cell-suspension to a sterile 15-ml tube.

2. Centrifuge the samples at $100 \times g$ for 10 min.

3. Discard the supernatant, and resuspend the spheres in 10 ml of PBS to wash off medium.

4. Centrifuge the samples at $100 \times g$ for 10 min, discard the supernatant and resuspend the pellet gently in 1 ml of 4% paraformaldehyde in PBS for 30 min at room temperature.

5. Rinse the neurospheres 3× with 10-ml washes of PBS, with centrifugation at $100 \times g$ for 10 min between each wash.

6. NSC-like neurospheres can be stored in PBS for several weeks prior to immunostaining.

3.4.3. Neural Differentiation of MSCs on Coverslips

1. Cells are harvested as normal from culture flasks during MSC- or NSC-like neurosphere passage, and plated at 1×10^5 cells/well onto poly-L-lysine coated coverslips.

2. Incubate the cells in neural induction medium containing either rh BDNF (which promotes neuronal differentiation) or rh PDGF-BB (promoting glial development).

3. After 7–10 days fix cells by removing medium and adding 4% paraformaldehyde for 30 min at 37°C. Remove paraformaldehyde and rinse coverslips with several washes of PBS. Coverslips can be stored in PBS for several weeks prior to immunostaining if required.

3.5. Immunostaining

3.5.1. Coverslip-Grown Cultures

1. If paraformaldehyde has been washed off previously, go to **step 2**. Otherwise rinse the coverslips with 3× washes of PBS, aspirating off PBS between washes.

2. Block and permeabilise cells with a solution of 4% normal goat serum and 0.1% Triton-X100 made up in PBS. Incubate the coverslips at 37° for 60 min.

3. Wash the coverslips with 3× washes of PBS.

4. Dilute the desired antibodies (**Table 1**) in blocking solution and incubate the coverslips for 60 min at room temperature, or 4°C overnight.

5. Wash the coverslips with 3× washes of PBS.

6. Dilute the secondary antibodies in blocking solution and incubate for 60 min at room temperature.

7. Wash the coverslips with 3× washes of PBS.

8. Mount the coverslips by inverting onto microscope slides using Vectashield mounting medium containing DAPI to visualise nuclei.

9. Visualise cells under a fluorescence-enabled microscope (**Fig. 1c, d**).

3.5.2. NSC-Like Neurospheres in Suspension

1. If the paraformaldehyde has been washed off previously, go to **step 2**. Otherwise rinse the sphere suspensions with 2× washes of PBS, centrifuging at $100 \times g$ for 10 min between each wash. All the steps in this protocol are carried out in 15-ml tubes.

2. Block and permeabilise spheres with a solution of 4% normal goat serum and 0.1% Triton-X100 made up in PBS. Incubate the spheres at 37° for 60 min.

3. Wash the spheres with 2× washes of PBS, centrifuging at $100 \times g$ for 10 min between each wash.

4. Dilute the desired antibodies (**Table 1**) in blocking solution and incubate the neurospheres for 60 min at room temperature. Ensure that a minimum of 500 μl of solution is used in each tube.

5. Wash the spheres with 2× washes of PBS, centrifuging at $100 \times g$ for 10 min between each wash.

6. Dilute the secondary antibodies in blocking solution and incubate the neurospheres for 60 min at room temperature. Ensure that a minimum of 500 μl of solution is used in each tube.

7. Wash the spheres with 2× washes of PBS, centrifuging at $100 \times g$ for 10 min between each wash.

8. Resuspend the final pellet in 50 μl of Vectashield mounting medium containing DAPI to visualise nuclei.

9. Mount the spheres between spacers on a microscope slide. Do this by securing two coverslips to a slide, with a space of approximately half a coverslip length between them. Place the sphere suspension onto the slide between the spacers, and place a final coverslip on top. Seal with nail polish.

10. Visualise NSC-like neurospheres under a fluorescence-enabled microscope.

4. Notes

1. The handling of all human bone marrow and marrow-derived cultures should be carried out in a Class 2 bio-cabinet or laminar flow cabinet. Sterilise equipment by autoclaving or by spraying with 70% ethanol.

2. MSC growth medium should be sterilised through 0.22-μm pore size filter units.

3. Aspiration of media from flasks and tubes should be carried out using sterile glass Pasteur pipettes attached to a vacuum pump and waste collector (containing bleach or similarly appropriate lab disinfectant).

4. All cells are cultured or differentiated at $37°C/5\%$ CO_2 in a humidified atmosphere.

5. Up to 1×10^6 cells can be frozen per cryotube. A higher number of cells per 25-cm^2 TC flask after thawing will mean that cells are confluent faster and those cultures can be restocked quicker if needed.

6. Freezing NSC-like neurospheres follows the same protocol as described above for MSC cultures. Collect the neurospheres 3–4 days after they have been passaged into low-attachment TC plates or flasks for freezing.

References

1. Bjorklund, A. (2000) Cell replacement strategies for neurodegenerative disorders. Novartis Found Symp **231**: 7–15; discussion 16–20.

2. Cao, Q., Benton, R. L. & Whittemore, S. R. (2002) Stem cell repair of central nervous system injury. J Neurosci Res **68**: 501–510.

3. Park, K. I., Ourednik, J., Ourednik, V., Taylor, R. M., Aboody, K. S., Auguste, K. I., Lachyankar, M. B., Redmond, D. E. & Snyder, E. Y. (2002) Global gene and cell replacement strategies via stem cells. Gene Ther **9**: 613–624.

4. Weissman, I. L. (2000) Translating stem and progenitor cell biology to the clinic: barriers and opportunities. Science **287**: 1442–1446.

5. Clarke, D. & Frisen, J. (2001) Differentiation potential of adult stem cells. Curr Opin Genet Dev **11**: 575–580.

6. Poulsom, R., Alison, M. R., Forbes, S. J. & Wright, N. A. (2002) Adult stem cell plasticity. J Pathol **197**: 441–456.

7. Serafini, M. & Verfaillie, C. M. (2006) Pluripotency in adult stem cells: state of the art. Semin Reprod Med **24**: 379–388.

8. Preston, S. L., Alison, M. R., Forbes, S. J., Direkze, N. C., Poulsom, R. & Wright, N. A. (2003) The new stem cell biology: something for everyone. Mol Pathol **56**: 86–96.

9. Prockop, D. J. (2002) Adult stem cells gradually come of age. Nat Biotechnol 20: 791–792.

10. Scolding, N. (2001) New cells from old. Lancet 357: 329–330.

11. Chopp, M. & Li, Y. (2002) Treatment of neural injury with marrow stromal cells. Lancet Neurol 1: 92–100.

12. Jiang, Y., Jahagirdar, B. N., Reinhardt, R. L., Schwartz, R. E., Keene, C. D., Ortiz-Gonzalez, X. R., Reyes M., Lenvik, T., Lund, T., et al. (2002) Pluripotency of mesenchymal stem cells derived from adult marrow. Nature **418**: 41–49.

13. Krause, D. S., Theise, N. D., Collector, M. I., Henegariu, O., Hwang, S., Gardner, R., Neutzel S. & Sharkis, S. J. (2001)

Multi-organ, multi-lineage engraftment by a single bone marrow-derived stem cell. Cell **105**: 369–377.

14. Pittenger, M. F., Mackay, A. M., Beck, S. C., Jaiswal, R. K., Douglas, R., Mosca, J. D., Moorman, M. A., Simonetti, D. W., Craig S. & Marshak, D. R. (1999) Multilineage potential of adult human mesenchymal stem cells. Science **284**: 143–147.

15. Deng, W., Obrocka, M., Fischer, I. & Prockop, D. J. (2001) In vitro differentiation of human marrow stromal cells into early progenitors of neural cells by conditions that increase intracellular cyclic AMP. Biochem Biophys Res Commun **282**: 148–152.

16. Kim, B. J., Seo, J. H., Bubien, J. K. & Oh, Y. S. (2002) Differentiation of adult bone marrow stem cells into neuroprogenitor cells in vitro. Neuroreport **13**: 1185–1188.

17. Sanchez-Ramos, J., Song, S., Cardozo-Pelaez, F., Hazzi, C., Stedeford, T., Willing, A., Freeman, T. B., Saporta S., Janssen, W., et al. (2000) Adult bone marrow stromal cells differentiate into neural cells in vitro. Exp Neurol **164**: 247–256.

18. Woodbury, D., Schwarz, E. J., Prockop, D. J. & Black, I. B. (2000) Adult rat and human bone marrow stromal cells differentiate into neurons. J Neurosci Res **61**: 364–370.

19. Hermann, A., Gastl, R., Liebau, S., Popa, M. O., Fiedler, J., Boehm, B. O., Maisel, M., Lerche H., Schwarz J., et al. (2004) Efficient generation of neural stem cell-like cells from adult human bone marrow stromal cells. J Cell Sci **117**: 4411–4422.

20. Akiyama, Y., Radtke, C., Honmou, O. & Kocsis, J. D. (2002) Remyelination of the spinal cord following intravenous delivery of bone marrow cells. Glia **39**: 229–236.

21. Akiyama, Y., Radtke, C. & Kocsis, J. D. (2002) Remyelination of the rat spinal cord by transplantation of identified bone marrow stromal cells. J Neurosci **22**: 6623–6630.

22. Azizi, S. A., Stokes, D., Augelli, B. J., DiGirolamo, C. & Prockop, D. J. (1998) Engraftment and migration of human bone marrow stromal cells implanted in the brains of albino rats – similarities to astrocyte grafts. Proc Natl Acad Sci USA **95**: 3908–3913.

23. Brazelton, T. R., Rossi, F. M., Keshet, G. I. & Blau, H. M. (2000) From marrow to brain: expression of neuronal phenotypes in adult mice. Science **290**: 1775–1779.

Chapter 9

Umbilical Cord Blood Cells

Jennifer D. Newcomb, Alison E. Willing, and Paul R. Sanberg

Summary

The umbilical cord of a healthy neonate contains within it a multipotential treatment for a myriad of diseases and injuries. What was once tossed into the biohazard waste without a second thought is now known to be a goldmine of antigenically immature cells that rival the use of bone marrow for reconstitution of blood lineages. Umbilical cord blood (UCB) is emerging as an effective and feasible clinical treatment as its availability increases and benefits are realized. Basic science research has demonstrated a broad therapeutic capacity ranging from cell replacement to cell protection and anti-inflammation in a number of animal disease and injury models. UCB is easily obtained with no harm to infant or mother and can be stored at cryogenic temperatures with relatively little loss of cells upon thaw. The heterogeneous mononuclear fraction has been identified and characterized and transplanted both locally and systemically to treat animal models of stroke, myocardial infarction, Amytrophic Lateral Sclerosis, San Filippo, spinal cord injury, traumatic brain injury, and age-related neurodegeneration, among others. In the pages to follow, we share protocols for the identification and research use of the mononuclear cell fraction of UCB.

Key words: Umbilical cord blood, Hematopoietic progenitor cell, Mononuclear cell isolation, Immunocytochemistry, Flow cytometry, UCB transplantation

1. Introduction

Tight government restrictions, which limit the availability and use of fetal and embryonic tissue, have prompted the pursuit for alternative sources of stem/progenitor cells. Currently, the clinical use of hematopoietic cells (HPC) derived from bone marrow (BM) is the gold standard treatment for malignant and nonmalignant blood diseases after myeloablative therapy (1, 2). However, use of BM has a number of logistical disadvantages that keep it from widespread use including inability to store cryogenically,

Neil J. Scolding and David Gordon (eds.), *Methods in Molecular Biology, Neural Cell Transplantation, vol. 549*
DOI: 10.1007/978-1-60327-931-4_9, © Humana Press, a part of Springer Science + Business Media, LLC 2009

difficult and invasive collection methods, high incidence of graft-vs.-host-disease (GvHD), and stringent human leukocyte antigen (HLA)-matching requirements to prevent graft rejection. Research continues to reveal that umbilical cord blood (UCB) use can circumvent these complexities and has a potentially unrivaled degree of plasticity. What used to be a biological waste product is now seen as a source of therapeutic cells that have the capability to treat a number of diseases and injuries. New parents now have the option to take an active role in treating future illness much like a biological insurance policy. Although, much of the currently banked UCB lies in wait for science to catch up with its theoretical curative capability, since 1988 UCB has saved and extended the life of more than 60,000 people worldwide suffering from hematopoietic malignancies, marrow failure, and immunodeficiency disorders *(2–7)*. UCB treatment becoming a clinical standard will require more efficient storage, recovery and expansion techniques, and a more thorough understanding of its mechanisms of action to ensure patient safety. Additionally, with more than 4 million babies born every year in the US alone, a campaign to educate future parents of the benefits of banking their newborn's cord blood or donating it will make its availability limitless. Here we provide protocols for the research use of UCB cells from isolation and storage to transplantation into animal models of disease. Each protocol herein is a standard in our research group and has been extensively tested for validity and reliability.

2. Materials

2.1. Isolation and Cryopreservation of Cord Blood Mononuclear Cells

1. Sterile Phosphate Buffered Saline (PBS).
2. Lymphocyte Separation Medium (LSM, Ficoll-hypaque; Sigma).
3. RPMI-1640 (Sigma).
4. Dimethyl Sulfoxide (DMSO).
5. Fetal Bovine Serum (FBS).
6. Sterile 2 ml Cryovials (Nalgene).

2.2. Immunocyto-chemistry of Cultured Umbilical Cord Blood Cells

1. Dulbecco's Modified Eagle's medium (DMEM, Gibco).
2. FBS.
3. Gentamicin (Sigma).
4. 4% paraformaldehyde in 0.1 M PBS.
5. Appropriate normal serum.

6. Triton X100.

7. Primary antibodies of interest.

8. Appropriate secondary antibodies.

9. Vectashield mounting medium with DAPI (Vector Laboratories).

2.3. Multicolor Flow Cytometry

1. Cold sterile PBS.

2. Trypan Blue dye (Gibco).

3. Bovine serum albumin (BSA).

4. Human serum.

5. Fluorochrome-conjugated antibody(s) of interest.

6. DAPI (Sigma).

7. FLOWJO flow cytometric analysis software (Tree Star, Inc.).

8. LSRII Flow Cytometer (BD).

2.4. UCB Migration to Injured Brain Extracts

1. Liquid nitrogen.

2. Clear DMEM (Gibco).

3. 0.22-μm filters (Millipore).

4. FBS.

5. Gentamicin.

6. Low-adherence 6-well culture dishes (Corning).

7. Trypan Blue dye.

8. 96-well Chemotax® Chamber plate (Neuro Probe).

9. Cell viability assay (CellTiter-Glo Kit, Promega).

10. Automated plate reader (Synergy, BioTek).

2.5. UCB Cell Preparation and I.V. Transplantation

1. Sterile PBS pH 7.4.

2. 1 M HEPES (Sigma).

3. Deoxyribonuclease I (DNase; Sigma).

4. Trypan Blue dye.

5. Isoflurane (Henry Schein Co.).

6. 5-0 Silk suture.

7. 26-gauge and 31-gauge needles.

2.6. UCB Cell Preparation and Local Delivery Using Stereotaxis

1. Isolyte S, pH 7.4 (*see* **Note 1**).

2. Isoflurane.

3. Stereotaxic frame (Kopf).

4. 10-μl Hamilton syringe.

2.7. Organotypic Hippocampal Culture

1. DMEM.
2. Horse serum (Vector Laboratories).
3. Hank's Balanced Salt Solution (HBSS; Gibco).
4. Glucose.
5. Glutamine.
6. Glucose-free DMEM (Gibco).
7. Propidium iodide (PI; Pharmingen).

2.8. Flow Cytometry for Brain Homogenates

1. HBSS.
2. BSA.
3. EDTA (BDH).
4. DNase.
5. 40-μm nylon cell strainers (BD).
6. Percoll (Gibco).
7. PBS.
8. FBS.
9. Fc block™ (BD Pharmingen).
10. Fluorochrome-conjugated antibody(s) of interest.
11. LSRII or FACScaliber flow cytometers (BD).
12. FLOWJO software (Tree Star, Inc.).

3. Methods

Unlike BM, and other HPC sources, UCB cells retain high viability after being kept at cryogenic temperatures until need arises. There are many techniques for volume reduction and cryopreserving these cells *(8–17)*, however, for FDA approval of a cell therapy the volume reduction and MNC isolation must be done in a closed system, free of operator error and contamination *(18–20)*. For laboratory research purposes, the UCB mononuclear cells are sufficiently prepared in an open or semi-closed system using density gradient separation. The cells can then be frozen in cryoprotectant and stored in liquid nitrogen.

3.1. Isolation and Cryopreservation of Cord Blood Mononuclear Cells MNC Isolation and Cryopreservation

1. Add 10 ml of sterile PBS to 25-ml fresh UCB in a 50-ml conical centrifuge tube and then invert the tube several times to mix.
2. Carefully underlay each tube with 12 ml LSM and centrifuge at $400 \times g$ for 30 min.

3. After centrifugation the mononuclear cell layer should be easily distinguishable at the interface of plasma and Ficoll. Remove the plasma.

4. Using a 10-ml sterile automatic pipette remove the mononuclear layer to just above the red cell layer.

5. Place the cells in a sterile 50-ml conical centrifuge tube and add RPMI-1640 to 45 ml (approximately a 1:2 dilution).

6. Centrifuge at $400 \times g$ for 15 min.

7. Decant the supernatant being careful not to disturb the cell pellet.

8. Consolidate all of the tubes into one 50-ml conical centrifuge tube and add RPMI-1640 to 45 ml.

9. Centrifuge at $400 \times g$ for 15 min.

10. Decant supernatant and transfer cell pellet to a 15-ml conical centrifuge tube and add RPMI-1640 to 14 ml. Mix gently by inversion.

11. Centrifuge at $400 \times g$ for 10 min.

12. Discard supernatant and add 2-ml RPMI-1640. Mix carefully using a sterile transfer pipette and transfer to a sterile 5-ml cryotube.

13. Prepare the suspension of cord blood at 10^6 cells/ml with 10% DMSO and FBS by slowly adding the freezing medium to the tube while gently shaking.

14. Aliquot 2 ml of the cell suspension to sterile cryovials and store in liquid nitrogen (*see* **Note 2**).

Lymphocytes and monocytes make up the majority of the mononuclear fraction (MNF) of cells in UCB *(21)*. Compared to peripheral blood (PB), UCB has a similar B-lymphocyte population and a lower absolute number of T-lymphocytes (CD3[+]) but a higher CD4[+]/CD8[+] ratio *(21, 22)*, as well as, a greater number of NK cells and fewer CD56[+] cytotoxic T-lymphocytes *(23)*. Additionally, UCB has fewer mature memory T-lymphocytes (CD45RO[+]) and higher proportions of young CD45RA[+] cells compared to adult sources *(22, 23)*. This lack of antigenicity contributes to UCB's reduced incidence of GvHD and less stringent donor-recipient matching requirements, leading to a shorter waiting period for treatment *(24, 25)*. A more abundant, antigenically naïve, population of hematopoietic stem/progenitor cells found in UCB led to enthusiasm in the neuron replacement therapy research field *(26–32)*. Not only does the MNF contain roughly 1% CD34[+] cells, a marker designated for its role in early hematopoiesis, but also these cells appear to be more primitive than those found in BM *(33, 34)*. Another subset of CD34[+] cells found in relatively high numbers in UCB are the more primitive CD133[+] cells. CD133[+] cells have been identified in fetal brain and in this area are considered to be neural stem cells (NSC) *(25, 35)*.

A nonhematopoietic stem cell, the mesenchymal stem cell (MSC), though sparse in comparison to BM, has also been identified in UCB *(36, 37)*. The MSC can give rise to such diverse phenotypes as osteoblasts, chondroblasts, adipocytes, hematopoietic, and neural cells (astrocytes and neurons) *(38, 39)*. Identifying MSCs is challenging because there is no universal consensus of its phenotypic description. In the literature, MSCs have been defined as being CD13, CD29, CD44, and CD90 positive and CD14, CD31, CD34, CD45, CD51/61, CD64, CD106, and HLA-DR negative *(37)* as well as CD73, CD90, CD105, and CD166 positive and CD31, CD34, CD45, CD80, and HLA-DR negative *(40)*. Agreement on the phenotype of these cells needs to first be reached before their true abundance in UCB can be known. Regardless, there is a general consensus that there are far fewer MSCs in UCB compared to BM *(41)*.

There are a number of ways to characterize the cells within UCB. Our group routinely uses culture methods to label individual cells immunocytochemically (**Table 1**). Flow cytometry is also an important tool in characterizing single cell suspensions. This technology can analyze multiple cell properties very quickly and provide objective quantitative results which compliment immunocytochemical methods.

3.2. Immunocyto-chemistry of Cultured Umbilical Cord Blood Cells Identification

1. Thaw cryopreserved UCB cells rapidly in a 37°C water bath (*see* **Note 3**).

2. Suspend the cells in DMEM with 10% FBS and 0.1% Gentamicin.

3. Plate cells in either 4-well plates or 100-mm dishes at a density of 100,000 cells/cm^2.

4. Cells can then be incubated at 37°C in 5% CO_2. If the cells are kept for longer than 4 days they will need fresh medium added.

5. Prior to immunolabeling, fix the cultures with 4% paraformaldehyde in 0.1 M PBS for 10 min and then wash with PBS three times.

6. Block the fixed cells with the appropriate 10% normal serum and 1% Triton X100 in PBS for 1 h at room temperature.

7. Incubate the cells in primary antibody for 24 h at 4°C (for a list of our commonly used antibodies and their working dilutions *see* **Table 1**).

8. Wash the cells three times for 10 min in PBS.

9. Incubate in the appropriate secondary antibody for 2 h at room temperature.

10. Nuclei can be visualized by counterstaining the nucleus using Vectashield mounting medium with DAPI.

Table 1
Immunocytochemistry

Antibody	Raised in	manufacturer	Dilution	Notes
CD3	Mouse	BD Pharmingen	1:50	T-lymphocytes
CD4	Mouse	BD Pharmingen	1:50	Helper/inducer T-lymphocytes
CD8	Mouse	BD Pharmingen	1:50	Cytotoxic/suppressor T-lymphocytes
CD11b	Mouse	BD Pharmingen	1:50	Granulocytes, monocytes, NK cells
CD14	Mouse	BD Pharmingen	1:50	Monocytes
CD19	Mouse	BD Pharmingen	1:50	B-lymphocytes
CD34	Mouse	BD Pharmingen	1:50	Hematopoietic progenitors
CD44	Mouse	BD Pharmingen	1:50	Hematopoietic cellular adhesion molecule
CD45	Mouse	BD Pharmingen	1:50	Common leukocytes
CD117	Mouse	BD Pharmingen	1:50	Progenitor cells
CD133	Mouse	BD Pharmingen	1:50	Stem cells
Human nuclei (hNuc)	Mouse	Chemicon	1:30	Nuclei
Human mito-chondria	Mouse	Chemicon	1:30	Mitochondria
TUJ1	Rabbit	Covance	1:1000	Early neuronal marker
MAP2	Rabbit	Chemicon	1:1000	Neuronal cytoskeleton protein
GFAP	Rabbit	Dako	1:500	Astrocytes
Nestin	Mouse	BD Transduction	1:100	Early neuronal marker
Ox6 (MHC II)	Mouse	Serotec	1:300	Activated microglia

3.3. Multicolor
Flow Cytometry
(See Note 4)

1. Thaw cells rapidly at 37°C and wash two times in cold sterile PBS.

2. Determine viability using Trypan Blue dye exclusion method.

3. Adjust volume to 10^6 cells per 100 μl buffer (cold sterile PBS with 1% BSA) in 1.5-ml microcentrifuge tubes.

4. Block Fc receptors with 10% human serum in PBS for 20 min.

5. Add fluorochrome-conjugated antibodies (*see* **Note 5**) and incubate for 30 min.

6. Wash cells two times with cold sterile PBS and centrifuge at $150 \times g$ for 5 min.

7. Following incubation with antibodies, add 75 μl of 0.35 μg/ml DAPI solution to every 300 μl of cells (approximately 10^6 cells, scale volume up or down as needed).

8. Acquisition can be done on a LSRII flow cytometer.

9. Scatter plots and percentages of fluorescent events can be generated using FLOWJO flow cytometric analysis software.

10. Using the scatter plots, gate cells based on their morphological characteristics.

Because of its large population of HPC, UCB was thought to be the ideal source of cells to be used for replacement of dead and/or diseased cells in a number of injuries and diseases. In vivo research has found that human UCB can ameliorate behavioral and physiological consequences in a number of animal disease models, the most exciting of which included diseases and injury of the brain *(42–54)*. Our lab and others have demonstrated UCB's growth as a multidimensional treatment. The exact mechanisms of action are still unclear but what is apparent is that UCB cells mold their utility to fit their need. New research in our lab has found that UCB may not only act as a cell replacement source, but also as a neurotrophic, neuroprotective, and anti-inflammatory agent. The most notable example of the multifaceted therapeutic effects of UCB is in the MCAO model of embolic stroke. In this rat model, we have achieved profound behavioral improvement and infarct reduction. Stroke involves a complicated cascade of inflammatory events which eventually lead to a pronounced area of cell death adjacent to the blocked vasculature. The inflammatory response to stroke injury has a distinct temporal nature and UCB cells given I.V. 48 h following stroke onset can reverse the injury. The temporal administration of UCB most likely modulates the immune/inflammatory response both peripherally and locally *(43–45)*.

In the complex, highly dynamic living model, neuroprotection and anti-inflammation effects of UCB treatment is difficult to tease apart. We have used an organotypic hippocampal culture system to attempt to define UCB's neuroprotective potential *(45)*. Inflammatory effects can be studied through ELISA (IL-10, TNF-α, IL-4, IL-1β, and IFNγ) on peripheral blood and homogenates of peripheral immune organs. Flow cytometry can also be done on single cell suspensions of brain and organ recovered from animal models *(45)*. Furthermore, at the immunohistochemical level, we can visualize UCB's treatment effects using the inflammatory markers GFAP, Ox6 (MCH II), OX42, and CD11b (*see* **Table 1** for our antibody dilutions).

3.4. UCB Migration to Injured Brain Extracts Inflammation and Neuroprotection

3.4.1. Preparation of Brain Tissue Extracts (See Note 6)

1. Remove brain rapidly and flash-freeze in liquid nitrogen (Store at −80°C until needed).
2. Partially thaw brain and dissect out regions of interest (e.g., hippocampus, striatum, and/or cortex).
3. Add 1 ml of clear DMEM for every 150 mg of tissue in a 50-ml conical centrifuge tube and homogenize (*see* **Note 7**).
4. Centrifuge homogenates at $400 \times g$ for 20 min.
5. Collect the supernatant and filter (can be stored at −80°C until needed).

3.4.2. Preparation of UCB Cells

1. Rapidly thaw cryopreserved UCB cells in a 37°C water bath and add to a 15-ml conical centrifuge tube containing 9 ml of clear DMEM, 5% FBS, and 1 μl/1 ml Gentamicin.
2. Centrifuge at $400 \times g$ for 15 min.
3. Remove supernatant and resuspend cells in 1 ml of media.
4. Plate cells in low-adherence 6-well culture dishes and incubate for 24 h at 37°C and 5% CO_2.
5. After 24 h, lift cells by gently pipetting and place in a 15-ml conical centrifuge tube.
6. Centrifuge at $400 \times g$ for 15 min and resuspend in 1 ml of media without FBS.
7. Determine viability using the Trypan Blue dye exclusion method (we only use cells with >80% viability).
8. Adjust cell concentration to 62,000 cells per 25 μl media.

3.4.3. Migration Assay

1. Pipette 300 μl of tissue extract (supernatant) into the bottom wells of a 96-well Chemotax® Chamber plate with a 5-μm pore size in triplicate.
2. Pipette UCB cells (62,000 cells per 25 μl) into the top well.
3. Incubate migration chamber at 37°C and 5% CO_2 for 4 h.
4. After incubation, remove top plate and centrifuge bottom plate at $300 \times g$ for 10 min so that all migrating cells will be forced to the bottom.
5. Remove half of the media (150 μl) and determine number of migrated cells using a cell viability assay (*see* **Note 8**).
6. The migration plate can then be read on an automated plate reader set to luminescence.

Despite our reasonable appreciation of the basic biology of UCB cells we still lack understanding of their dynamics in the living organism. Phenotype means very little if the cells do not function appropriately. Buzanska's group *(55)* have shown that the floating population of proliferating cells (CD34⁻/CD45⁻) have functioning sodium/potassium channels which respond to exposure to dopamine, serotonin, γ-aminobutyric acid, and acetylcholine. This finding further fueled enthusiasm that UCB could become a viable option as a source of transplantable cells that would replace diseased/dead cells. However, the culture dish is far removed from the unique complexities of the biological animal so the next step is to establish that these cells function in vivo. Our group transplanted UCB cells into the subventricular zone of the rat neonate and demonstrated that a handful of cells did differentiate into neurons and glia *(32)*.

Although it is still unclear whether UCB cells behave the same as phenotypically identical endogenous cells, we do recognize they have an inimitable therapeutic utility beyond replacement of lost cells. UCB cells can ameliorate both motor asymmetry and physiological consequences brought about by middle cerebral artery occlusion in a rat model of stroke *(42–47)*. Early studies demonstrated that some of the transplanted cells expressed the neural antigens NeuN, MAP2, and GFAP indicating that the stroke injury may have led these cells down a neural lineage *(42)*. Interestingly, UCB cells delivered intravenously (I.V.) were more effective in improving long-term outcome than when delivered directly into the affected brain parenchyma postulating an effect on the peripheral immune response to the stroke *(46)*. And, though infiltration of the I.V. administered UCB cells into the brain was minimal, in vitro studies suggest that injured brain does produce chemotactic signals that attract specific cells to the site of injury *(56)*.

Until recently, UCB cell infusion in our and others' animal stroke models was arbitrarily set to 24 h following injury. The results from the aforementioned migration assay led to a surprising discovery. UCB cells appeared to have a definite temporal nature to their migration, which in striatal extracts (injured area in MCAO rat models of stroke), was as late as 72 h *(56)*. The clear temporally dependent nature of migration was discovered and led to affirmation that when UCB cells are administered I.V. at a specific time point postinjury onset maximum recovery results in vivo *(43)*. In many cases, stroked animals did not display physiological or behavioral consequences from their injury. It is believed that the peak of the inflammatory response to the injury coincides with the optimal therapeutic window. Moreover, temporal I.V. administration of UCB cells is a more critical determinant of recovery than cell dose based on both behavioral and physiological measures of recovery *(43, 44)*.

Our group employs I.V. delivery of UCB cells in other animal models using the penile, tail, jugular, and femoral veins. Choice of method often depends on ease of administration. In studies where only male animals are used, the penile vein is fast and efficient, requiring minimal anesthesia time and no surgical cut down. Jugular vein delivery is employed in studies utilizing both male and female mice. And, though, I.V. is currently the delivery route of choice in our group, local administration is still required to answer questions of the cells' direct effect on the brain/injured area.

3.5. UCB Cell Preparation and I.V. Transplantation In Vivo

3.5.1. UCB Cell Preparation

1. Rapidly thaw cryopreserved UCB cells in a 37°C water bath.
2. Transfer thawed cells from cryovial to a 15-ml conical centrifuge tube containing buffer [9.8 ml sterile PBS pH 7.4, 200 µl 1 M HEPES and 500 µl DNase (5 mg/ml)].
3. Centrifuge at room temperature at $200 \times g$ for 10 min.
4. Remove supernatant and resuspend cells in 10-ml buffer.
5. Repeat **step 4**.
6. Resuspend cell pellet in 1-ml buffer and determine viability using the Trypan Blue dye exclusion method.
7. Determine cell density needed for injection and prepare desired number of cells in buffer.

3.5.2. Femoral Vein Infusion

1. Anesthetize animal with 3% Isoflurane in O_2 at 2 l/min.
2. Shave inside thigh area and prepare using alternating betadine and alcohol scrubs.
3. Using blunt dissection, isolate the femoral vein.
4. Using 5-0 silk suture put a tie on the distal vein to prevent cells from flowing back.
5. Introduce a 26-gauge needle attached to a 1-ml syringe containing cell suspension into the lumen of the vein.
6. Deliver cells (usually 100–500 µl) over a 5-min period.
7. Withdraw the needle and place an additional suture proximal to the hole to prevent bleeding and cells from leaking out.
8. Close the incision.

3.5.3. Tail Vein Infusion

1. Anesthetize animal as described above.
2. Apply a hot compress to the base of the tail to dilate the vein (or run under hot water, taking care not to scald the skin).
3. Introduce a 26-gauge needle attached to a 1-ml syringe containing the cell suspension into the lumen of the vein and deliver cells as described above.
4. Withdraw needle and apply pressure to the point of entry to prevent bleeding.

3.5.4. Jugular Vein Infusion

1. Anesthetize animal and prepare area as described above.
2. Using blunt dissection, isolate the jugular vein.

3. Ligate the vein as described in the femoral vein infusion methods and make a small hole in the vessel using a 26-gauge needle.

4. Insert a 31-gauge needle attached to a 10-μl Hamilton syringe containing the cell suspension into the hole and suture into place.

5. Deliver the cells over a 5-min period.

6. Withdraw needle and tighten suture to avoid bleeding and loss of cells and close incision.

3.5.5. Penile Vein Infusion

1. Anesthetize animal as described above.

2. Expose penile vein and clean area with alcohol.

3. Introduce a 26-gauge needle attached to a 1-ml syringe containing cell suspension into the lumen of the vein and deliver cells over a 5-min period.

4. Withdraw needle and apply pressure to entry site to avoid bleeding and cell loss.

3.6. UCB Cell Preparation and Local Delivery Using Stereotaxis

3.6.1. Cell Preparation for Local Cell Delivery

1. Thaw cryopreserved cells rapidly in a 37°C water bath.

2. Wash in Isolyte S, pH 7.4 and centrifuge cells three times (200 × g for 10 min).

3. Determine viability and adjust to desired cell concentration (we use a cell concentration of 100,000 cells/μl).

4. Aliquot the suspension into volumes slightly larger than the final desired volume to be delivered (*see* **Note 9**).

3.6.2. Preparation of Animal for Local Cell Delivery

1. Animals are anesthetized with 3% Isoflurane in O_2 at 2 l/min and placed in a stereotaxic frame.

2. Shave the head and surgically prepare using alternating betadine and alcohol scrubs.

3. Make a midline incision with a scalpel and clean away the facia from the skull.

4. Make all medial/lateral (ML) and anterior/posterior (AP) measurements from the Bregma landmark and dorsal/ventral (DV) measurements from dura (*see* **Note 10**).

5. Bore a small hole in the skull with a drill taking care not to drill past the lower extent of the skull.

6. With a fine dental pick, carefully place a small hole in the dura for the needle to pass through (*see* **Note 11**).

7. Load a 10-μl Hamilton syringe with the cell suspension just prior to delivery to prevent cells from clumping.

8. Lower the needle into place at the desired coordinates and wait for 2 min to allow the tissue to decompress around the needle.

9. Deliver cells over the course of the next 5 min. Leave the needle in place for an additional 5 min prior to withdrawing slowly (*see* **Note 12**).

10. Close incision with wound clips.

3.7. Organotypic Hippocampal Culture

1. Isolate the hippocampi and cut into coronal slices (1,000-mm thick).

2. Transfer slices into 24 mm-diameter membrane inserts and place into 24-well culture trays with 1.5 ml of medium, consisting of 50% DMEM, 25% heat-in-activated horse serum, and 25% HBSS supplemented with 4.5 mg/ml glucose and 1 mM glutamine, per well.

3. Change the media every 2 days for 14 days.

4. Place slices in glucose-free medium and subject to 45 min of hypoxia in a hypoxia chamber (a hypoxic environment is produced in a humidified chamber at 37°C that is prepared by flushing at 8 l/min for 15 min with 1% oxygen, 5% CO_2, and 94% nitrogen).

5. Following hypoxia, culture an additional 3 days under ischemic conditions (deoxygenated, glucose-free DMEM).

6. For the co-culture group, place the membrane inserts with the ischemic hippocampal slices into a well with UCB cells (50,000 cells/ml). For the non-co-culture group, place the hippocampal slices into cell-free media (1 ml per well).

7. Following culture period, stain the cells with PI to identify the dying cells.

8. Delineate Hippocampal areas CA1, CA2, and CA3 and the dentate gyrus (DG) and quantify the optical density using a density analyzer program (Image J, NIH, *see* **Note 13**).

3.8. Flow Cytometry for Brain Homogenates

1. Remove the brain rapidly, separate the hemispheres and place into ice-cold HBSS containing 0.2% BSA, 0.01 mol/l EDTA, and 10 mg/ml DNase.

2. Homogenize the brain tissue and passed through a 40-μm nylon cell strainer.

3. Centrifuge the suspension two times at $400 \times g$ for 10 min at room temperature.

4. Resuspend the pellet in 70% Percoll in BS.

5. Overlay suspension with a gradient containing 30% Percoll solution in PBS and centrifuge at $600 \times g$ for 45 min at room temperature.

6. Collect the cells from the Percoll to PBS interface and wash once with HBSS with 10% FBS.

7. Resuspend cells in flow buffer (0.02% BSA, 0.05 mM EDTA in PBS) containing Fc block™ to inhibit nonspecific binding.

8. Incubate with appropriate fluorochrome-conjugated antibody(s) (*see* **Note 1**) for 60 min in the dark with and on ice.

9. After incubation with antibody, wash two times with flow cytometry buffer.

Acquisition can be performed on the LSRII or FACScaliber flow cytometers and analysis done using FLOWJO software.

4. Conclusion

All of the research described within this chapter follows a common thread; it attempts to define UCB's role in recovery from disease and injury. Clearly, UCB works uniquely depending on its environment and the nature of the injury; a presumed molding to fit the need. The multifaceted nature of UCB's therapeutic effects is no doubt a function of its heterogenous make up, and also is what eludes researchers attempting to classify its benefits. Of course, this is the goal and the necessity before UCB can achieve pervasive use in the clinic, both putting to rest the question of safety and appropriately matching its utility to fit the specific illness. The therapeutic potential of UCB is far reaching from common ailments to exotic diagnoses. Any disease with an inflammatory component or need of neuroprotection may benefit from UCB cell treatment. If discovering its mechanisms of action to ensure safety in its use and perfecting the expansion process is all that remains, then its widespread penetration into hospitals and clinics is inevitable.

5. Notes

1. While Isolyte S, pH 7.4 was the vehicle we used in the initial transplant studies, PBS with HEPE and DNase as described in **Subheading 3.5.1** works equally well.

2. For best results, the cells should be gradually cooled to −80°C in a freezing container (Nalgene) before placing in liquid nitrogen.

3. The vial of cells is only left in the water bath until approximately 95% of the suspension is thawed to minimize the shock of thawing to the cells.

4. To maintain cell viability it is crucial that cells be kept cold (preferably on ice) throughout blocking, antibody incubations, and until acquisition on a Flow cytometer. Ideally, samples should be run immediately following **step** 7 of the

multicolor Flow Cytometry protocol, but if not practical, can be stored on ice in the dark for up to 2 h.

5. Specific concentration of antibody should be determined by titration. Titration is especially important for multiple color flow cytometry where even a small change in buffer volume can dilute an antibody to nonsaturating levels.

6. Brain tissue extracts were harvested from rats that underwent middle cerebral artery occlusion (MCAO). For MCAO description, refer to Newcomb et al. *(43)*.

7. It is critical that the media is colorless; otherwise phenol red may interfere with the luminescence readings for cell viability.

8. While there is more than one way to measure cell viability, we have found the Promega CellTiter-Glo kit works well in this application.

9. The cell suspension is divided into single transplant aliquots for two reasons. First, there is less likelihood of contamination and second, cell viability is maintained longer as the suspension is not being continually mixed and remixed for multiple transplants. Aliquot volume should be greater than the final volume that is to be delivered, allowing for needle/syringe deadspace and testing of delivery prior to transplant.

10. Stereotactic coordinates commonly used in our group. (a) Rats. *Striatum* coordinates (delivered into two adjacent sites in our studies): AP 1.2 mm, ML 2.7 mm, DV –5.2 mm, and –4.7 mm. Tooth bar set at zero. *Subventricular zone*: AP 1.6 mm, ML ±1.5 mm, DV –4.2 mm. Tooth bar set at –2.3. *Hippocampus*: AP –3.8 mm, ML ±2.5 mm, DV –1.5 mm. Tooth bar set at –2.3. (b) Mice. *Ventricle*: AP 3.8 mm, ML ±0.8 mm, DV 2.0 mm. Tooth bar set at zero.

11. By piercing the dura in advance, there is less damage to cortex from tearing of the dura or compression as the needle moves into the tissue.

12. Waiting 5 min prior to withdrawing the needle allows the cell suspension to diffuse away from the needle tip, decreasing the likelihood of pulling the transplanted cells up and out along the needle tract when the needle is withdrawn.

13. Image J software for image processing and analysis can be downloaded from the NIH Web site: http://rsb.info.nih.gov/ij/

Acknowledgments

The authors would like to thank Dr. Svitlana Garbuzova-Davis, Dr. Ning Chen, Dr. Mary Newman, Dr. Nicole Kuzmin-Nichols, and Dr. Martina Vendrame for contributing their protocols.

Many of the UCB MNCs used in the above protocols were obtained from Saneron CCEL Therapeutics, Inc. PRS is a cofounder and AEW is a consultant to Saneron CCEL Therapeutics, Inc.

References

1. Armitage, J. (1994). Bone marrow transplantation. *New Eng. J. Med.* **330**, 827–838.

2. El-Badri, N.S., Kazi, Z., Sanberg, P.R. (2004). Stem cell transplantation for hematologic malignancies. *Cell Transplant.* **13**, 721–723.

3. Barker, J.N., Weisdorf, D.J., DeFor, T.E., Blazar, B.R., McGlave, P.B., Miller, J.S., Verfaillie, C.M., Wagner, J.E. (2004). Transplantation of 2 partially HLA-matched umbilical cord blood units to enhance engraftment in adults with hematologic malignancy. *Blood* **105**, 1343–1347.

4. Gluckman, F., Rocha, V. (2005). History of the clinical use of umbilical cord blood hematopoietic cells. *Cytotherapy* **7**, 219–227.

5. Laughlin, M.J., Barker, J., Bambach, B., Koc, O.N., Rizzieri, D.A., Wagner, J.E., Gerson, S.L., Lazarus, H.M., Cairo, M., Stevens, C.E., Rubinstein, P., Kurtzberg, J. (2001). Hematopoietic engraftment and survival in adult recipients of umbilical cord blood from unrelated donors. *New Eng. J. Med.* **344**, 1815–1822.

6. Tse, W., Laughlin, M.J. (2005). Umbilical cord blood transplantation: a new alternative option. *Hematol.* 2005(1), 377–383.

7. Wagner, J.E., Broxmeyer, H.E., Byrd, R.L., Zehnbauer, B., Schmeckpeper, B., Shah, N., Griffin, C., Emanuel, P.D., Zuckerman, K.S., Cooper, S., Carow, C., Bias, W., Santos, G.W. (1992). Transplantation of umbilical cord blood after myeloablative therapy: analysis of engraftment. *Blood* **79**, 1874–1881.

8. Escolar, M.L., Poe, M.D., Provenzale, J.M., Richards, K.C., Allison, J., Wood, S., Wenger, D.A., Pietryga, D., Wall, D., Champagne, M., Morse, R., Krivit, W., Kurtzberg, J. (2005). Transplantation of umbilical-cord blood in babies with infantile Krabbe's disease. *N. Eng. J. Med.* **352**, 2069–2081.

9. Fujisaki, G., Kami, M., Kishi, Y. (2004). Cord-blood transplants from unrelated donor in Hurler's syndrome. *N. Eng. J. Med.* **351**, 506–507.

10. NMDP (1999). Cord Blood Bank Standard Operating Procedures. National Marrow Donor Program & National Heart, Lung, and Blood Institute.

11. Kögler, G., Sarnowski, A., Wernet, P. (1998). Volume reduction of cord blood by Hetastarch for long-term stem cell banking. *Bone Marrow Transplant.* **22**(Suppl 1), S14–S15.

12. Rubinstein, P., Dobrila, L., Rosenfield, R.E., Adamson, J.W., Migliaccio, G., Migliaccio, A.R., Taylor, P.E., Stevens, C.E. (1995). Processing and cryopreservation of placental/umbilical cord blood for unrelated bone marrow reconstitution. *Proc. Natl Acad. Sci. USA* **92**(22), 10119–10122.

13. Perutelli, P., Catellani, S., Scarso, L., Cornaglia-Ferraris, P., Dini, G. (1999). Processing of human cord blood by three different procedures for red blood cell depletion and mononuclear cell recovery. *Vox Sang.* **76**(4), 237–240.

14. Schwinger, W., Benesch, M., Lackner, H., Kerbl, R., Walcher, M., Urban, C. (1999). Comparison of different methods for separation and *ex vivo* expansion of cord blood progenitor cells. *Ann. Hematol.* **78**(8), 364–370.

15. Almici, C., Carlo-Stella, C., Mangoni, L., Garau, D., Cottafavi, L., Ventura, A., Armanetti, M., Wagner, J.E., Rizzoli, V. (1995). Density separation of umbilical cord blood and recovery of hemopoietic progenitor cells: implications for cord blood banking. *Stem Cells* **13**(5), 533–540.

16. Tsang, K.S., Li, K., Huang, D.P., Wong, A.P., Leung, Y., Lau, T.T., Chang, A.M., Li, C.K., Fok, T.F., Yuen, P.M. (2001). Dextran sedimentation in a semi-closed system for the clinical banking of umbilical cord blood. *Transfusion* **41**(3), 344–352.

17. Harris, D.T., Schumacher, M.J., Rychlik, S., Booth, A., Acevedo, A., Rubenstein, P., Bard, J., Boyse, E.A. (1994). Collection, separation and cryopreservation of umbilical cord blood for use in transplantation. *Bone Marrow Transplant.* **13**(2), 135–143.

18. Dracker, R.A. (1996). Cord blood stem cells: how to get them and what to do with them. *J. Hematother.* **5**(2), 145–148.

19. Reems, J.A., Fujita, D., Tyler, T., Moldwin, R., Smith, S.D. (1999). Obtaining an accepted investigational new drug application to operate an umbilical cord blood bank. *Transfusion* **39**(4), 357–363.

20. Zuck, T.F. (1996). The applicability of cGMP to cord blood cell banking. *J. Hematother.* **5**(2), 135–137.

21. Pranke, P., Failace, R.R., Allebrandt, W.F., Steibel, G., Schmidt, F., Nardi, N.B. (2001). Hematologic and immunophenotypic characterization of human umbilical cord blood. *Acta Haematol.* **105**, 71–76.

22. Harris, D., Schumacher, M., Locascio, J., Besencon, F., Olson, G., DeLuca, D., Shenker, L., Bard, J., Boyse, E. (1992). Phenotypic and functional immaturity of human umbilical cord blood T lymphocytes. *Proc. Natl Acad. Sci. USA* **89**, 10006–10010.

23. D'Arena, G., Musto, P., Cascavilla, N., Di Giorgio, G., Fusili, S., Zendoli, F., Carotenuto, M. (1998). Flow cytometric characterization of human umbilical cord blood lymphocytes: immunophenotypic features. *Haematologica* **83**, 197–203.

24. Rocha, V., Cornish, J., Sievers, E.L., Filipovich, A., Locatelli, F., Peters, C., Remberger, M., Michel, G., Arcese, W., Dallorso, S., Tiedemann, K., Busca, A., Chan, K.W., Kato, S., Ortega, J., Vowels, M., Zander, A., Souillet, G., Oakill, A., Woolfrey, A., Pay, A.L., Green, A., Garnier, F., Ionescu, I., Wernet, P., Sirchia, G., Rubinstein, P., Chevret, S., Gluckman, E. (2001). Comparison of outcomes of unrelated bone marrow and umbilical cord blood transplants in children with acute leukemia. *Blood* **97**, 2962–2971.

25. Tamaki, S., Eckert, K., He, D., Sutton, R., Doshe, M., Jain, G., Tushinski, R., Reitsma, M., Harris, B., Tsukamoto, M., Gage, F., Weissman, I., Uchida, N. (2002). Engraftment of sorted/expanded human central nervous system stem cells from fetal brain. *J. Neurosci. Res.* **69**, 976–986.

26. Bender, J.G., Unverzagt, K.L., Walker, D.E., Lee, W., Van Epps, D.E., Smith, D.H., Stewart, C.C., To, L.B. (1991). Identification and comparison of CD34-positive cells and their subpopulations from normal peripheral blood and bone marrow using multicolor flow cytometry. *Blood* **77**(12), 2591–2596.

27. Bu a ska, L., Machaj, E.K., Zablocka, B., Podja, Z., Doma ska-Janik, K. (2002). Human cord blood-derived cells attain neuronal and glial features in vitro. *J. Cell Sci.* **115**, 2131–2138.

28. Chen, N., Hudson, J.E., Walczak, P., Misiuta, I., Garbuzova-Davis, S., Jiang, L., Sanchez-Ramos, J., Sanberg, P.R., Zigova, T., Willing, A.E. (2005). Human umbilical cord blood progenitors: the potential of these hematopoietic cells to become neural. *Stem Cells* **23**(10), 1560–1570.

29. Jurga, M., Markiewicz, I., Sarnowska, A., Habich, A., Kozlowska, H., Lukomska, B., Bu a ska, L., Doma ska-Janik, K. (2006). Neurogenic potential of human umbilical cord blood: neural-like stem cells depend on previous long-term culture conditions. *J. Neurosci. Res.* **83**, 627–637.

30. Sanchez-Ramos, J.R., Song, S., Kamath, S.G., Zigova, T., Willing, A., Cardozo-Pelaez, F., Stedeford, T., Chopp, M., Sanberg, P.R. (2001). Expression of neural markers in human umbilical cord blood. *Exp. Neurol.* **171**, 109–115.

31. Wu, A.G., Michejda, M., Mazumder, A., Meehan, K.R., Menendez, F.A., Tchabo, J.G., Slack, R., Johnson, M.P., Bellanti, J.A. (1999). Analysis and characterization of hematopoietic progenitor cells from fetal bone marrow adult bone marrow peripheral blood, and cord blood. *Pediatr. Res.* **46**(2), 163–169.

32. Zigova, T., Song, S., Willing, A.E., Hudson, J.E., Newman, M.B., Saporta, S., Sanchez-Ramos, J., Sanberg, P.R. (2002). Human umbilical cord blood cells express neural antigens after transplantation into the developing rat brain. *Cell Transplant.* **11**(3), 265–274.

33. Cardosa, A.A., Li, M.L., Batard, P., Hatzfeld, A., Brown, E.L., Levesque, J.P., Sookdeo, H., Panterne, B., Sansilvestri, P., Clark, S.C., Hatzfeld, J. (1993). Released from quiescence of CD34+CD38– human umbilical cord blood cells reveal their potentiality to engraft adults. *Proc. Natl Acad. Sci. USA* **90**, 8707–8711.

34. Conrad, P., Emerson, S.G. (1998). Ex vivo expansion of hematopoietic cells from umbilical cord blood for clinical transplantation. *J. Leuk. Biol.* **64**, 147–155.

35. Uchida, N., Buck, D.W., He, D., Reitsma, M.J., Masek, M., Phan, T.V., Tsukamoto, A.S., Gage, F.H., Weissman, I.L. (2000). Direct isolation of human central nervous system stem cells. *Proc. Natl. Acad. Sci. USA* **97**, 14720–14725.

36. Goodwin, H.S., Bicknese, A.R., Chien, S.N., Bogucki, B.D., Quinn, C.O., Wall, D.A. (2001). Multilineage differentiation activity by cells isolated from umbilical cord blood: expression of bone, fat, and neural markers. *Biol. Blood Marrow Transplant.* **7**: 581–588.

37. Yang, S.-E., Ha, C.-W., Jung, M.H., Jin, H.-J., Lee, M.K., Song, H.S., Choi, S.J., Oh, W., Yang, Y.-S. (2004). Mesenchymal stem/progenitor cells developed in cultures from UC blood. *Cytotherapy* **6**, 476–486.

38. Javazon, E.H., Beggs, K.J., Flake, A.W. (2004). Mesenchymal stem cells: paradoxes of passaging. *Exp. Hematol.* **32**, 414–425.

39. Kögler, G., Sensken, S., Airey, J.A., Trapp, T., Müschen, M., Feldhahn, N., Liedtke, S., Sorg, R.V., Fischer, J., Rosenbaum, C., Greschat, S., Knipper, A., Bender, J., Degistirici, Ö., Gao, J., Caplan, A.I., Colletti, E.J., Almeida-Porada, G., Müller, H.W., Zanjani, E., Wernet, P. (2004). A new human somatic stem cell from placental cord blood with intrinsic pluripotent differentiation potential. *J. Exp. Med.* **200**, 123–135.

40. Robinson, S., Niu, T., de Lima, M., Ng, J., Yang, H., McMannis, J., Karandish, S., Sadeghi, T., Fu, P., de Angel, M., O'Connor, S., Champlin, R., Shpall, E. (2005). Ex vivo expansion of umbilical cord blood. *Cytotherapy* 7, 243–250.

41. Wexler, S.A., Donaldson, C., Denning-Kendall, P., Rice, C., Hows, J.M. (2003). Adult bone marrow is a rich source of human mesenchymal 'stem' cells but umbilical cord blood and mobilized adult blood are not. *Br. J. Haematol.* **121**, 368–374.

42. Chen, J., Sanberg, P.R., Li, Y., Wang, L., Lu, M., Willing, A.E., Sanchez-Ramos, J., Chopp, M. (2001). Intravenous administration of human umbilical cord blood reduces behavioral deficits after stroke in rats. *Stroke* **32**, 2682–2688.

43. Newcomb, J.D., Ajmo, Jr. C.T., Sanberg, C.D., Sanberg, P.R., Pennypacker, K.R., Willing, A.E. (2006). Timing of cord blood treatment after experimental stroke determines therapeutic efficacy. *Cell Transplant.* **15**, 213–223.

44. Vendrame, M., Cassady, J., Newcomb, J., Butler, T., Pennypacker, K.R., Zigova, T., Davis Sanberg, C., Sanberg, P.R., Willing, A.E. (2004). Infusion of human umbilical cord blood cells in a rat model of stroke dose-dependency rescues behavioral deficits and reduces infarct volume. *Stroke* **35**, 2390–2395.

45. Vendrame, M., Gemma, C., de Mesquita, D., Collier, L., Bickford, P.C., Davis Sanberg, C., Sanberg, P.R., Pennypacker, K.R., Willing, A.E. (2005). Anti-inflammatory effects of human cord blood cells in a rat model of stroke. *Stem Cells Dev.* **14**, 595–604.

46. Willing, A.E., Lixian, J., Milliken, M., Poulos, S., Zigova, T., Song, S., Hart, C., Sanchez-Ramos, J., Sanberg, P.R. (2003). Intravenous versus intrastriatal cord blood administration in a rodent model of stroke. *J. Neurosci. Res.* **73**(3), 296–307.

47. Willing, A.E., Vendrame, M., Mallery, J., Cassady, C.J., Davis, C.D., Sanchez-Ramos, J., Sanberg, P.R. (2003) Mobilized peripheral blood cells administered intravenously produce functional recovery in stroke. *Cell Transplant.* **12**, 449–454.

48. Chen, R., Ende, N. (2000). The potential for the use of mononuclear cells from human umbilical cord blood in the treatment of amyotrophic lateral sclerosis in SOD1 mice. *J. Med.* **31**, 21–30.

49. Garbuzova-Davis, S., Willing, A.E., Zigova, T., Saporta, S., Justen, E.B., Lane, J.C., Hudson, J.E., Chen, N., Davis, C.D., Sanberg, P.R. (2003). Intravenous administration of human umbilical cord blood cells in a mouse model of amyotrophic lateral sclerosis: distribution, migration and differentiation. *J. Hematother. Stem Cell Res.* **12**, 255–270.

50. Garbuzova-Davis, S., Willing, A.E., Desjarlais, T., Davis Sanberg, C., Sanberg, P.R. (2005).Transplantation of human umbilical cord blood cells benefits an animal model of Sanfilippo syndrome type B. *Stem Cells Dev.* **14**, 384–394.

51. Garbuzova-Davis, S., Gografe, S.J., Davis Sanberg, C., Willing, A.E., Saporta, S., Cameron, D.F., Desjarlais, T., Daily, J., Kuzmin-Nichols, N., Chamizo, W., Klasko, S.K., Sanberg, P.R. (2006). Maternal transplantation of human umbilical cord blood cells provides prenatal therapy in Sanfilipo type B mouse model. *FASEB J.* **20**, 485–487.

52. Henning, R.J., Abu-Ali, H., Balis, J.U., Morgan, M.B., Willing, A.E., Sanberg, P.R. (2004).Human umbilical cord blood mononuclear cells for the treatment of acute myocardial infarction. *Cell Transplant.* **13**, 729–739.

53. Lu, D., Sanberg, P.R., Mahmood, A., Li, Y., Wang, L., Sanchez-Ramos, J., Chopp, M. (2002). Intravenous administration of human umbilical cord blood reduces neurological deficit in the rat after traumatic brain injury. *Cell Transplant.* **11**, 275–281.

54. Saporta, S., Kim, J.-J., Willing, A.E., Fu, E.S., Davis, C.D., Sanberg, P.R. (2003). Human umbilical cord blood stem cells infusion in spinal cord injury: engraftment and beneficial influence on behavior. *J. Hematother. Stem Cell Res.* **12**, 271–278.

55. Sun, W., Buzanska, L., Domanska-Janik, K., Salvi, R.J., Stachowiakc, M.K. (2005). Voltage-sensitive and ligand-gated channels in differentiating neural stem-like cells derived from the nonhematopoietic fraction of human umbilical cord blood. *Stem Cells* **23**, 931–945.

56. Newman, M.B., Willing, A.E., Mansrea, J.J., Davis-Sanberg, C., Sanberg, P.R. (2005). Stroke-induced migration of human umbilical cord blood cells: time course and cytokines. *Stem Cells Dev.* **14**(5), 576–586.

Chapter 10

Animal Models of Neurodegenerative Diseases

Wendy Phillips, Andrew Michell, Harald Pruess, and Roger A. Barker

Summary

Animal models of neurodegenerative disease are excellent tools for studying pathogenesis and therapies including cellular transplantation. In this chapter, we describe different models of Huntington's disease and Parkinson's disease, stereotactic surgery (used in creation of lesion models and transplantation) and finally transplantation studies in these models.

Key words: Animal models, Neurodegenerative disease, Transplantation, Neural precursor cells, Huntington's disease, Parkinson's disease

1. Introduction

Huntington's disease (HD) is a devastating inherited neurodegenerative condition characterised by movement disorders, cognitive deficits leading to dementia and psychiatric dysfunction. The genetic defect leads to a mutation in a ubiquitous protein, huntingtin, and neuronal loss, particularly in the caudate nucleus in early disease. The prevalence of the disease is around 1 in 10,000 *(1)*. Onset is usually in middle age, although the range of disease onset is wide. The cell and molecular biology of the disease has been extensively studied in recent years and this has lead to potential therapies although none so far have been shown clinically to slow disease progression. The genetic abnormality in HD is an expanded triplet repeat of the bases: cysteine, adenine and guanine (CAG) (which encode glutamine, Q) on exon 1 of the "interesting transcript" (IT15) gene on chromosome 4 *(2)*.

Parkinson's disease (PD) is a common neurodegenerative disease with a prevalence of 1 in 800; classically patients exhibit

Neil J. Scolding and David Gordon (eds.), *Methods in Molecular Biology, Neural Cell Transplantation, vol. 549*
DOI: 10.1007/978-1-60327-931-4_10, © Humana Press, a part of Springer Science + Business Media, LLC 2009

bradykinesia, rigidity and tremor as well as cognitive and psychiatric disturbance, but the disease is heterogeneous, affecting patients in a wide variety of ways *(3)*. At the heart of the disease, pathologically, is degeneration of dopaminergic cells within the substantia nigra pars compacta *(4)*, with characteristic intracytoplasmic eosinophilic inclusions called Lewy bodies. The resulting reduction in dopaminergic input to the striatum is thought to be responsible especially for the bradykinesia seen in the disease, although symptoms are only noted once about 50% of the nigral cells are lost *(4)*.

Animal models are invaluable in the study of early disease, of pathophysiological mechanisms (and in the case of HD, for example, models have been useful in elucidating the role of normal wild-type huntingtin), in interventional studies, and in therapeutic screens including cell therapy and transplantation. However, it must be realised that they still remain as models of the disease and as such only approximate to the clinical scenario of patients and these disorders.

We will describe the different models of HD and PD, how the lesion models are created and how to choose between these different models. We will then describe what factors need to be considered when embarking upon transplantation studies, including those employing stem cells.

2. Materials

2.1. Striatal Excitotoxic Lesion

1. Quinolinic acid (Sigma). Dissolve first in 5 μl of 1 M NaOH and then dilute in 0.1 M phosphate buffered saline (to pH 7.4).

2. Phosphate buffered saline: 8.5 g NaCl, 0.4 g di-hydrogen sodium phosphate, 1 g di-sodium hydrogen phosphate, 1 l dH$_2$O, pH to 7.4 with HCl.

2.2. The 6-Hydroxydopamine Rodent Model

1. 0.5 μl of 6-OHDA (5.5 mg free base/ml in 0.1% ascorbate, 0.9% saline) is injected over 2 min into the right medial forebrain bundle at the following stereotactic coordinates: –2.0 mm anterior to bregma, –0.7 mm lateral to bregma and –4.8 mm vertical below the dura.

2. Intraperitoneal *met*-amphetamine sulphate: 2.5 mg/kg dissolved in 0.9% saline. Net ipsilateral rotations (rightward rotational bias, turns/min) are recorded over 90 min.

2.3. Transplantation Studies

1. Dulbecco's modified Eagle's media; Gibco/Life Technologies; Paisley, Scotland.

2. BrdU: Roche, Susssex, UK.

3. Rat anti-BrdU: 1 in 50 (Accurate Chemicals) or 1 in 500 (Serotec). For BrdU immunolabelling, sections should be incubated in 2 M hydrochloric acid (HCl) at 37°C for 30 min, then washed three times for 10 min in 0.1 M borate buffer (38.1 g Borax, Sigma; 1 l distilled water, pH 8.5 with NaOH) prior to block.

4. Acrylic matrix brain slicer; World Precision Instruments, Inc., Sarasota, FL, USA.

5. Papain (Lorne Laboratories, Twyford, UK), Dispase (Boehringer Mannheim, UK), DNase (Lorne Laboratories).

6. Proliferation media: Neurobasal medium (Gibco/Life Technologies) with 2% B27 (Gibco/Life Technologies), 1% PSF (Gibco/Life Technologies), 1% L-glutamine, 20 ng/ml FGF-heparin (R&D Systems, Oxford, UK), 20 ng/ml EGF (Sigma, Poole, UK).

7. Polybrene (1,5-dimethyl-1,5-diazaundecamethylene polymethobromide): Sigma-Aldrich, Germany.

8. X-Gal staining kit GALS-1KT, Sigma (St. Louis, MO, USA)

3. Methods

3.1. Animal Models of HD

3.1.1. Excitotoxic Models

Given that HD is caused by a single gene defect, many robust genetic animal models have been developed. Before this defect was discovered in 1993 (2), animal models were primarily excitotoxic, targeting the site of earliest pathology in HD, namely the striatum. Intrastriatal excitotoxins in particular kainic acid, was the first to be used (5) but has now been generally superseded by quinolinic acid, which more faithfully reflects HD pathology, with loss of medium spiny gamma-amino-butyric-acid-ergic projection neurons and sparing of NADPH diaphorase neurons (6) (Fig. 1). The mitochondrial toxin 3-nitropropionic acid given intraperitoneally produces selective striatal toxicity and a model of HD (7). As is often the case with animal models, the discovery that systemic mitochondrial inhibitors produces selective striatal toxicity not only yielded a valuable model of disease, but gave clues as to the nature of the pathophysiology of HD and fuelled the "excitotoxic theory" of HD. There is certainly some evidence for impaired energy metabolism in HD, and degeneration of mitochondria, although this may be a secondary phenomenon (8–12).

The use of stereotactic apparatus allows for the injection of numerous substances including excitotoxins for lesioning and cells for transplantation. Stereotactic (Gk "movement in space") operations use three-dimensional co-ordinates to locate a specific target in the brain. Stereotactic frames are available for small

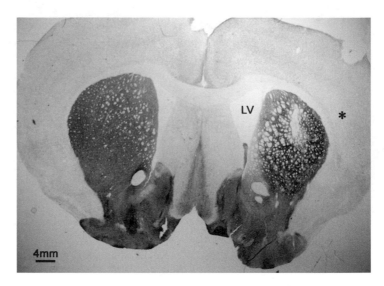

Fig. 1. Photomicrograph of a sagittal section through a mouse brain showing a lesion produced by stereotactic injection of quinolinic acid into the left striatum. The striatum has been identified using acetylcholinesterase immunohistochemistry. *Asterisk* indicates the side of the injection; *LV* lateral ventricle.

animals and can be used to create specific targeted lesions; in the case of HD, the striatum.

1. Fill a Hamilton syringe with the desired substance, e.g. quinolinic acid. Quinolinic acid (Sigma) is first dissolved in 1 M NaOH, then diluted in 0.1 M phosphate buffered saline (pH 7.4). Quinolinic acid has been used in mice in doses of 30 and 40 nmol, e.g. *(13, 14)*.

2. Prepare a sterile field with sterile scissors, forceps, surgical clamps, sutures, scalpel and swabs.

3. Place the animal on a heat pad, positioned underneath a woolen blanket. Anaesthetise the animal (*see* **Note 1**), ensuring of course, the anaesthetic gas is directed to the animal and the gas scavenging system is on. The animal is often given antibiotic and analgesia (*see* **Note 2**). Position the head in the frame (**Fig. 2**).

4. Note the position of the tooth bar (most often set at zero), and insert the teeth onto the tooth bar (*see* **Note 3**).

5. The head is then kept still by securing two metal bars (different sizes of course for different animals) on either side of the head (*see* **Note 4**), and the position of the bars should be noted using the read-off scale on the bottom of the apparatus. The head should be straight and level, and firmly positioned so as not to slip when tapped from above. A small incision (around 1 cm) can now be made in the scalp, after painting with iodide.

6. Using a large microscope and lamp, bregma (the junction of the sagittal and coronal sutures) should be located (**Fig. 3**, *see* **Note 5**).

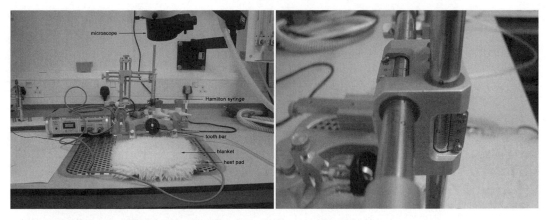

Fig. 2. Stereotactic apparatus. (*Left*) The frame used to hold the animal's head is pictured behind the woolen blanket and heat pad. The arm that holds the Hamilton syringe and drill can be seen above the frame. The anaesthetic gas and scavenging system is located at the black oval mask, above which sits the tooth bar, and the ear-bars can be seen on either side. (*Right*) Detail of the co-ordinate arms.

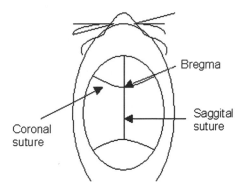

Fig. 3. Bregma is located between the sagittal and coronal sutures and is used as a neutral point for stereotactic co-ordinate readings.

Bregma can then be described using co-ordinates by bringing the drill over bregma and then noting the position of the arm of the drill holder by reading off three sets of co-ordinates for its position in space (**Fig. 2**, *see* **Note 6**).

7. Move the scalp skin laterally to accommodate the position of the drill, otherwise hair and skin will become caught when the drill starts. Once the drill is positioned correctly, with the drill (*see* **Note 7**) located just above the skull, it can be operated using a footpad, and lowered carefully onto the skull to create a small burr hole. Then slowly raise and remove the drill (*see* **Note 8**).

8. Replace the drill with a Hamilton syringe pre-filled with the appropriate substance. The substance, for example quinolinic acid, should be injected slowly (say 1 μl over 2–3 min), the syringe left in place for another few minutes (the same amount

of time for each injection) to prevent reflux up the needle and then slowly removed (*see* **Note 9**).

9. Suture the incision, remove the animal from the frame and leave it to recover in a heated padded box with regular checks to ensure that the mouse recovers, wakes and becomes active (*see* **Note 10**).

3.1.2. Genetic Mouse Models of HD

The discovery of the genetic defect in 1993 led to the creation of genetic models of HD, with a transgenic mouse created in 1996 by Gil Bates and colleagues in London *(15)*. Genetic models are arguably superior to excitotoxic models in HD, in that they reflect the disease process more faithfully, the mice can be genetically manipulated and cross-bred to tease out the molecular pathogenesis of the disease, and also because the striatum is not the only structure to dysfunction and degenerate in HD *(16, 17)*. However, they are also not without limitations.

Transgenic Mouse Models

Transgenic models are created by the random insertion of genetic material into the host genome. Thus, the host retains endogenous expression of the gene in question and the inserted genetic material is under the control of an exogenous promoter and not in the appropriate genomic context. In the case of HD, the genetic material inserted comprises either exon-1 or the full-length huntingtin gene.

There are now many transgenic models of HD, which differ in background, fragment/full-length gene, CAG repeat length, promoter, and endogenous expression of the transgene. These differences produce very diverse pathologies (in terms of protein aggregates and inclusions, cell loss, neuronal dysfunction) and phenotypes. In general, few repeats, low expression and full-length transgenes produce a model with a less severe disease course. Conversely, many repeats, high expression and N-terminal fragments of the gene produce models with severe disease and a short lifespan. Despite this, there is very little striatal cell loss in these models, which may limit their usefulness *(15)*. It has been argued that these latter models reflect juvenile HD, given the large number of repeats and severe phenotype *(1, 18)*. Genetic background also influences mouse phenotype, with for example, different rates of neurogenesis *(19)* and response to excitotoxins *(20)* dependent on strain. Even for the same model, such as the R6/2 model *(15)*, breeding procedures and husbandry are different in different laboratories, which may significantly affect the life expectancy and behaviour of the mice *(21)*. Due to genetic anticipation, whereby trinucleotide repeats expand in length as they are inherited possibly due to gametic instability *(22)*, the CAG repeat length expands with every generation. If this is not checked and mice are not re-derived (i.e. not using males from late generations with very expanded repeat lengths) then CAG repeat lengths can reach large proportions, into the 200s. In humans, the htt gene normally has less than 36 CAG repeats and the mutant gene has greater than 39 CAG repeats *(1, 23, 24)*. Having said this, however, it may be litter origin and environmental sources of variability (such as age at weaning, environmental enrichment, provision of mash to prevent dehydration, number of cage mates, etc.) that matter more to phenotype *(21)*. For example, the Morton colony of R6/2 mice is environmentally enriched, with running wheels and toys as well as regular provision of wet mash, and the mice from this colony tend to have greater longevity and less severe disease than mice from non-enriched colonies *(25)*. These differences may affect some experimental results given that environmental enrichment can affect neurogenesis, for example *(26)*.

Knock-In Models

Knock-in models are created by the insertion of CAG repeats or CAG repeats with a part of the gene (from an exogenous source; these are known as chimeric knock-ins) directly into the relevant host gene. Thus, the inserted genetic material will be under the endogenous promoter and in the appropriate genomic context. Initially it was thought that these models were poor because they displayed little pathological and phenotypical changes but this view has been revised with more knowledge about these mice and it may be that they in fact are the best model of adult onset HD precisely because the disease onset is slow *(27)*. The chimeric knock-in mice may be better models because they carry a sequence that encodes for a polyproline tract, which may be involved in huntingtin protein structure and some protein–protein interactions *(27)*.

3.1.3. Non-Mouse Models of HD

Cellular models of HD allow high throughput screening of therapeutically promising compounds, as well as allowing detailed study of molecular mechanisms of pathogenesis mediated by the mutant gene. Cellular models comprise, for example, cells cultured from HD patients, such as fibroblasts *(28)* or cells transfected with the mutant gene *(29)*.

Neuronal intranuclear inclusions and neurodegeneration have been described in a *Drosophila melanogaster* model of HD *(30)*, and also a *Caenorhabditis elegans* model *(31)*. These models are important as an intermediate step from cellular to mammalian models, because of their simplicity and availability of a complete genome sequence. They are often used for suppressor screens, which aim to identify genes that modify the disease process.

Zebrafish (*Dania rerio*) are attractive model organisms by virtue of their rapid developmental time and transparency of their embryos, and have been used successfully in mutagenesis screens. The zebrafish HD gene homologue has been recently characterised and a zebrafish HD model developed *(32, 33)*.

The choice of model for a particular experiment should depend on the pathology (cell loss, abnormal protein aggregation and inclusions), neuronal dysfunction and transcriptional changes, and the motor and behavioural phenotype displayed by the mouse. Mice with a short lifespan allow rapid completion of experiments but have less value in exploring specifically the gradual disease process. Mice that have been extensively studied, such as the R6/2 model, are useful experimentally. Finally, logistical matters such as commercial and inter-laboratory availability must also be considered.

To answer a specific scientific question, however, it is best that many models are studied. For example, some HD models (the fragment models) are resistant to excitotoxins (13, 14, 34, 35), some display no change from wild-type animals (36) and others show enhanced sensitivity (37).

There is an extensive literature on mouse models of HD but details on specific models are beyond the scope of this chapter. There are some excellent and comprehensive review articles on this subject (38, 39) and specifically on the R6/2 model (40), use of models in elucidating pathophysiology (41) and therapeutics, (40) and knock-in models (27).

3.2. Animal Models of PD

There are a large number of animal models of PD, from transgenic *Drosophila*, which are particularly useful for genetic research, to *C. elegans* models and up to primate models of disease (42, 43). Perhaps the most widely used are rodent models – on the one hand transgenic models (which we shall illustrate by using the example of α-synuclein transgenic mice) and on the other hand toxic models (we shall give the example of 6-hydroxydopamine (6-OHDA) lesioned rats since they are often used to assess the efficacy of intracerebral transplants). There exist other interesting models, such as those that use adenoviral vectors to drive targeted overexpression of α-synuclein in adult animals (44). Detailed descriptions of these are beyond the scope of this chapter.

3.2.1. The 6-Hydroxydopamine Rodent Model

Although there are several toxin-induced models of PD, the 6-OHDA model (**Fig. 4**) will be explained here since it can be used to quantify the response to striatal transplants of dopamine-rich tissue. 6-OHDA is structurally extremely similar to dopamine (with an extra hydroxyl group on the carbon ring) and is actively taken up by dopaminergic and noradrenergic transporters on dopaminergic cells, thus explaining its relative selectivity for this cell type. Within the neuron it is metabolised to create free radicals, including hydrogen peroxide and reactive quinines, which damage molecules, resulting in cell death.

Since 6-OHDA does not cross the blood–brain barrier it must be injected directly into the brain to exert its effects. Injection into the medial forebrain bundle, that carries ascending

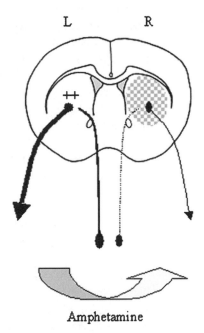

Fig. 4. The 6-OHDA model of Parkinson's disease. Right-sided injection of the medial forebrain bundle causes degeneration of the nigrostriatal tract on that side. Amphetamine induces dopamine release from the unlesioned dopaminergic striatal nerve terminals, increasing rotational bias towards the side with the lesion. Transplantation of dopaminergic cells to the right striatum can ameliorate the rotational asymmetry if >200 dopaminergic cells survive. Apomorphine rotation relies on using low dose apomorphine to stimulate supersensitive dopamine receptors in the dopamine-denervated striatum. In this case, the animal rotates away from the side of the lesion.

dopaminergic and serotonergic fibres to the striatum, results in irreversible dopaminergic cell death in the substantia nigra and ventral tegmentum. Unlike Parkinson's disease, there are no Lewy bodies formed, but like in the human disease there does seem to be relative sensitivity of dopaminergic neurons in the substantia nigra pars compacta. Bilateral lesions result in severe aphagia, adipsia and akinesia *(45)*, whereas unilateral lesions result in a far less sick animal with contralateral deficits in grip strength, sensory neglect and a postural bias towards the lesioned side *(46)*. Rodents receiving unilateral injections of this toxin develop a spontaneous head-to-tail turning motion in the direction of the lesion, a behaviour exacerbated by stimulation of the dopaminergic system, such as use of an ice-cold surface, or pharmacologically with amphetamine or apomorphine.

Importantly, the lesions induced by 6-OHDA injection of the medial forebrain bundle cause cell death within 24 h, and are stable and complete within a couple of weeks *(46)*. This is in contrast to the lesions induced by intrastriatal injection of the toxin that produces a progressive retrograde loss of dopaminergic neurons that can last for months *(47)*, and is helpful if chronic slow degenerative lesions are required.

The assessment of unilateral medial forebrain bundle 6-OHDA lesions (and their subsequent repair) is most commonly performed after intraperitoneal injection of amphetamine, and measured using automated rotometer bowls *(48)*. Amphetamine acts for 3–4 h, causing catecholamine release, especially dopamine, from pre-synaptic nerve terminals on the intact side, thus increasing the tendency for mice to rotate towards the side of the lesion (**Fig. 4**). This technique provides a rotation score that is stable, and, most importantly, one that can be quantified. The rate of turning correlates with striatal dopamine depletion *(46)*, and the effects of grafting can be examined animal by animal *(49)*. Furthermore, the technique benefits from being extremely well studied and relatively well understood. There are potential pitfalls with the interpretation of results from this technique. First, an improved rotation score, in other words a reduction in rotational bias, does not only arise from a restoration of dopaminergic tone on the lesioned side since a further lesion on the intact side would have a similar effect. Furthermore, it has even been shown that a large striatal lesion ipsilateral to the original 6-OHDA lesion can result in improved rotational behaviour, especially after apomorphine, which relies on its action on the supersensitive dopamine receptors in the striatum on the side of the median forebrain bundle lesion *(50)*. Thus to better understand the benefit of any new treatment (such as transplantation), a range of behavioural tests should be considered, such as paw reaching, stepping and measurement of sensory neglect in addition to rotational behaviour *(51)*.

3.2.2. α-Synuclein Transgenic Mice

The first transgenic mouse overexpressing wild-type human α-synuclein was reported in 2000 *(52)*. Expression of α-synuclein was driven by the platelet-derived growth factor-β promoter, resulting in prominent nuclear and cytoplasmic α-synuclein-positive granular inclusions in several regions of the brain including the substantia nigra, but no fibrillar pathology characteristic of Lewy bodies. The mice lost dopaminergic nerve terminals, and the striatal tyrosine hydroxylase (TH) levels and activity were reduced in line with the highest expression of the transgene. In time, a reduction in rotarod locomotor ability was detectable.

Since this original publication, numerous other α-synuclein transgenic mouse lines have been studied. Of course each is different to the other, but a few general points can be made. First, it is interesting to note that models using the TH promoter to target α-synuclein expression to the nigra fail to cause dopaminergic cell loss, suggesting that the relationship between α-synuclein accumulation and dopaminergic cell death is not straightforward.

Second, some of the mouse lines that overexpress the A53T mutant, but not A30P, tend to develop a striking motor neuron degeneration in the spinal cord, accompanied by the denervation of neuromuscular synapses *(53–55)*. Third, although not all publications report such findings, there does seem to be an association between the finding of truncated α-synuclein and filamentous neuropathology in some transgenic mouse models *(54, 55)*.

There are many possible explanations for the differences in pathology and phenotype between these different transgenic mouse models. Of course, both the type of α-synuclein used and the promoter driving its expression are likely to affect the outcome – one suggestion has been that the transgene needs to be expressed in a variety of cell types, including glia, to have maximal effect *(56)*. The level of expression achieved in the models varies from about 0.5 to 30 times the normal amount of endogenous mouse α-synuclein, suggesting some models may fail to achieve a critical threshold. For example, expression levels in the nigra tend to be very low, whereas in some models there is high expression in the motor neurons of the spinal cord, explaining the sensitivity of these cells described above. Of course more practical details such as the background strain of mouse used and type of behavioural test used will affect the results and their comparability between different models.

Considering these transgenic mouse lines together, there are a number of reasons why mouse dopaminergic neurons might be relatively immune to the overexpression of mutant α-synuclein. Seven amino acids differ between human and mouse endogenous α-synuclein, including the encoding of threonine in place of alanine at position 53, perhaps explaining why A53T mutants are handled differently. Moreover, these subtle differences may affect the tertiary structure of α-synuclein, and therefore its interaction with the human species and aggregation. Alternatively, mouse neurons may activate cell survival pathways not available in other species, such as those that have been proposed to explain the minimal amyloid plaque formation in transgenic mouse models of Alzheimer's disease expressing mutant human amyloid precursor protein *(57)*.

Although many models demonstrated some dysfunction of the dopaminergic system (be it subtle alterations in neuronal morphology or decreased striatal dopamine concentrations), the lack of dopaminergic cell death is disappointing. It is possible that this may relate to the observation that the interaction of human α-synuclein with the endogenous mouse species results in slower fibrillization *(58)*, suggesting that an α-synuclein null background might create a more severe model (if the fibrils are the toxic species) – perhaps more akin to the profound model in *Drosophila*, that have no natural endogenous α-synuclein *(42)*.

3.3. Transplantation in HD and PD Models

Transplantation studies in HD and PD models have served to research and optimise this technique as a prelude to human studies of foetal transplantation, which are designed to repair primary pathology through cell replacement. These studies have also served other purposes, for example, xenotransplants as an alternative to foetal allo-transplantation have revealed the processes underlying rejection and immunology, and neural precursor cell (NPC) transplants also as an alternative to foetal allo-transplantation and have been used to study the behaviour of stem cells when grafted into different microenvironments.

When attempting transplantation studies, several factors must be considered, such as donor tissue, host site and microenvironment, the transplantation procedure itself, post-graft "training" *(59)* and behavioural testing (**Fig. 5**).

3.3.1. Donor Tissue

In terms of donor tissue, the nature and number of cells to be transplanted are critical factors, along with their age of harvest and storage. In addition, the modes of engrafting including the media in which the cells are stored and to be transplanted are important factors. Dulbecco's modified Eagle's media has been found to be superior to Hank's balanced salt solution in the preparation of embryonic striatal tissue for example *(60)*. Furthermore, NPCs may survive better when grafted as neurospheres, rather than as a cell suspension *(61)*. Cells should be kept on ice and used as soon as possible. The number of cells transplanted is important *(62)*, and cells from embryos proliferate and survive better than cells from adult tissue *(63)*.

3.3.2. Labelling of Cells for Transplantation

When transplanting cells from one species into the brain of a recipient from the same species, the donor cells must be labelled in order to distinguish donor cells from host cells. If transplanting R6/2 cells into a wild-type brain for example, then the donor cells

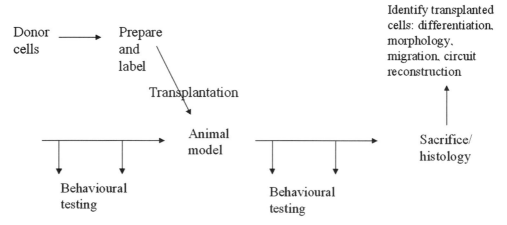

Fig. 5. Schema illustrating a standard protocol for assessment of transplanted tissue in a disease model.

can be easily identified using antibodies against mutant huntingtin *(64)*. Cells from a male donor can be distinguished from a female host by virtue of a Y chromosome assay *(65)*. Donor tissue may come from an animal expressing a fluorescent protein, but often the host and the donor should come from the same genetic background, which may not always be possible. Specifically using NPCs as donor tissue, these can be labelled using either bromo-deoxyuridine (BrdU) or viral transfection.

BrdU is a thymidine analogue and intercalates into DNA during the S-phase of mitosis thus labelling proliferating cells *(66)*. Proliferating NPCs in culture can be labelled with pulses of BrdU and a dose of 10 µM is generally accepted to be non-toxic to dividing cells. Caldwell and colleagues, however, have shown that this concentration of BrdU is selectively toxic to developing neurons, something that does not happen at a dose of 0.2 µM *(67)*. NPCs are often labelled 24 h before transplantation. At least the time of the S-phase should be given prior to transplantation; in the case of a mouse this is 6.5 h *(68)*, but adult dentate gyrus neurospheres are smaller and take more time to develop than subventricular zone neurospheres *(69)* and so it may be necessary to pulse for, say 3 days prior to transplantation (personal observations). It is wise to take an aliquot of cells prior to transplantation and confirm the labelling of cells in vitro using immunohistochemistry. The use of BrdU to label proliferating cells has been cautioned by a recent study showing that in adult (and developing) brains, host cells can take up BrdU from transplanted dead cells pre-labelled with 10 µM BrdU *(70)*.

Genetic modification of NPCs to express green fluorescent protein (GFP) is commonly done ex vivo using viral vectors. Several vectors are available for the gene delivery, each with their own advantages. The vast spectrum of retro-, adeno- and lentiviral vectors for gene-transfer is reviewed elsewhere *(71)*.

Lentiviral vectors are widespread due to their high transduction efficiency, and protocols for generation of a GFP-expressing lentiviral vectors are well established *(72)*. For example, deletions in the flanking long-terminal repeat sequence of the human immunodeficiency virus (HIV)-1-derived lentiviral vector cause self-inactivation of the vector but still allow integration. The GFP gene is subcloned into the construct and driven by the cytomegalovirus (CMV) promotor.

Retroviruses require integration into the host genome in order for gene expression to occur, and some retroviruses lack nuclear transport machinery and can enter the cell only when the nuclear membrane breaks down during cell division *(73)*. Retrovirus-mediated gene-transfer with a retroviral construct can include incorporation of an internal ribosomal entry site (IRES) as well as lacZ gene encoding *E. coli* β-galactosidase for later lacZ screening or XGal histochemistry *(74)*. In either case, the vector is then used for cotransfection of cells together with a helper plasmid encoding required lenti- or retroviral genes. The cells eventually become viral producer cells and release high titers of replication-incompetent virus into the medium. For infection of NPCs, virus aliquots are thawed on ice and applied onto dividing NPCs in the presence of polybrene. Medium is replenished after 4 h. After 3 days cells are washed, trypsinised, counted, diluted and used for transplantation. With multiple passages the number of GFP- or XGal-positive cells gradually decreases, thus cells should be studied during the first 4 weeks in culture.

Using viral gene-transfer further enables NPCs to produce a certain protein that might promote cell differentiation or survival after transplantation. Thus, viral transfection of NPCs can be a useful tool beyond cell tracking.

3.3.3. Host Site

The age of the host must be considered, and younger animals are able to better integrate grafts *(75)*. It may be the case that grafts integrate better into a lesioned, as opposed to intact, area *(76)*. Further, it is possible that transplanted cells survive better in the tissue surrounding a lesion rather than the lesion core. In the case of PD, tissue is grafted into the striatum rather than the substantia nigra, as it is difficult to produce new long distance dopaminergic projection neurons from the substantia nigra to the striatum.

3.3.4. Transplantation Procedure

Transplantation is performed using stereotactic apparatus. When using neurospheres for transplantation, they should either be small enough to fit into the needle bore without becoming stuck or they should be chopped prior to transplantation *(77)*. Chopping of neurospheres is very simple: intact spheres are picked up using a Pasteur pipette and placed into a petri dish containing a small volume of media. The neurospheres are then simply chopped into small pieces using a sterile razor blade. To ensure chopped neurospheres are actually taken up into the needle of the Hamilton syringe, this can be directly visualised under a microscope just prior to transplantation. It is often a good idea to use a needle with internal plunger to ensure material does not become stuck within the needle bore. Post-procedure, we usually allow at least 7 days for graft integration.

4. Notes

1. In our laboratory, for mice, surgery is often performed under halothane anaesthesia, induced in a chamber with inflow of 3-litres/minute (l/min) halothane (Fluothane, Merial) and

1.5-l/min oxygen. Maintenance anaesthesia is achieved with 2.5-l/min halothane, 0.6-l/min oxygen and 0.8-l/min nitrous oxide through a nose mask.

2. A subcutaneous injection of analgesic (Rimadyl, 5 mg/kg; Pfizer, Kent, UK is used in our laboratory) and antibiotic (Terramycin, 60 mg/kg; Pfizer) dissolved in 0.9% sterile saline may be given prior to the onset of surgery.

3. This can be fiddly and it helps to open the jaw slightly and slide the open mouth from the side of the tooth bar then position it in the centre of the bar.

4. The bars should be secured firmly to just avoid the head slipping, but not too tightly. It often helps to slide a finger under the animal's jaw to keep the head level while placing the bars. In our laboratory, for mice, we use bars with grips at the end, which is secured on either side of the head, although some laboratories use "ear-bars", where a tapered end is inserted into the animals' ears.

5. When the skull is moist, bregma can be seen more easily, and it often helps to tap gently on the skull, which dips it down slightly thus visually enhancing the juxtaposition of the sutures.

6. The location of the target structure will have been determined previously, using a brain atlas, previous literature and/or a previous experiment using, say, India ink to inject and locate a structure. For example, for the mouse striatum, the following co-ordinates have been used: 0.7 mm rostral to bregma, 1.9 mm lateral to the midline and 2.5 mm ventral to the dura *(13, 14)*. Thus, 0.7 mm should be added (if the desired position was caudal to bregma then of course the relevant amount should be subtracted) to the antero-posterior bregma reading and 1.9 mm should be added to the lateral bregma reading and the drill positioned accordingly **(Fig. 2)**.

7. For mice, a Foredom dental drill with a bit of 0.5-mm diameter is often used.

8. It is helpful to make a mark on the drill arm demonstrating which way to turn the knob to ensure the drill is raised and not erroneously lowered, which might inadvertently damage the cortex.

9. Again, it is helpful to have a mark on the drill arm to ensure the syringe is raised and not erroneously inserted more deeply.

10. This is remarkably quick; around 5–10 min for mice.

Acknowledgments

The authors would like to thank Mrs. Pam Tyers for excellent technical advice and assistance and Dr. Jenny Morton for provision of transgenic HD mice. The corresponding author's work was supported by the Medical Research Council.

References

1. Bates, G., Harper, P., and Jones, L. (2002) Huntington's Disease. 3rd edition, Oxford University Press, Oxford.
2. The Huntington's Disease Collaborative Research Group (1993) A novel gene containing a trinucleotide repeat that is expanded and unstable on Huntington's disease chromosomes *The Huntington's Disease Collaborative Research Group Cell* 72, 971–983.
3. Foltynie, T., Brayne, C., and Barker, R.A. (2002) The heterogeneity of idiopathic Parkinson's disease *J. Neurol.* 249, 138–145.
4. Fearnley, J.M. and Lees, A.J. (1991) Ageing and Parkinson's disease: substantia nigra regional selectivity *Brain* 114, 2283–2301.
5. Coyle, J.T. and Schwarcz, R. (1976) Lesion of striatal neurones with kainic acid provides a model for Huntington's chorea *Nature* 263, 244–246.
6. Beal, M.F., Ferrante, R.J., Swartz, K.J., and Kowall, N.W. (1991) Chronic quinolinic acid lesions in rats closely resemble Huntington's disease *J. Neurosci.* 11, 1649–1659.
7. Brouillet, E., Jacquard, C., Bizat, N., and Blum, D. (2005) 3-Nitropropionic acid: a mitochondrial toxin to uncover physiopathological mechanisms underlying striatal degeneration in Huntington's disease *J. Neurochem.* 95, 1521–1540.
8. Goebel, H.H., Heipertz, R., Scholz, W., Iqbal, K., and Tellez-Nagel, I. (1978) Juvenile Huntington chorea: clinical, ultrastructural, and biochemical studies *Neurology* 28, 23–31.
9. Jenkins, B.G., Koroshetz, W.J., Beal, M.F., and Rosen, B.R. (1993) Evidence for impairment of energy metabolism in vivo in Huntington's disease using localized 1H NMR spectroscopy *Neurology* 43, 2689–2695.
10. Tabrizi, S.J., Workman, J., Hart, P.E., Mangiarini, L., Mahal, A., Bates, G., et al. (2000) Mitochondrial dysfunction and free radical damage in the Huntington R6/2 transgenic mouse *Ann. Neurol.* 47, 80–86.
11. Guidetti, P., Charles, V., Chen, E.Y., Reddy, P.H., Kordower, J.H., Whetsell, W.O., Jr., et al. (2001) Early degenerative changes in transgenic mice expressing mutant huntingtin involve dendritic abnormalities but no impairment of mitochondrial energy production *Exp. Neurol.* 169, 340–350.
12. Brennan, W.A., Jr., Bird, E.D., and Aprille, J.R. (1985) Regional mitochondrial respiratory activity in Huntington's disease brain *J. Neurochem.* 44, 1948–1950.
13. Hansson, O., Petersen, A., Leist, M., Nicotera, P., Castilho, R.F., and Brundin, P. (1999) Transgenic mice expressing a Huntington's disease mutation are resistant to quinolinic acid-induced striatal excitotoxicity *Proc. Natl Acad. Sci. USA* 96, 8727–8732.
14. Phillips, W., Morton, A.J., and Barker, R.A. (2005) Abnormalities of neurogenesis in the R6/2 mouse model of Huntington's disease are attributable to the in vivo microenvironment *J. Neurosci.* 25, 11564–11576.
15. Mangiarini, L., Sathasivam, K., Seller, M., Cozens, B., Harper, A., Hetherington, C. et al. (1996) Exon 1 of the HD gene with an expanded CAG repeat is sufficient to cause a progressive neurological phenotype in transgenic mice *Cell* 87, 493–506.
16. Pavese, N., Andrews, T.C., Brooks, D.J., Ho, A.K., Rosser, A.E., Barker, R.A., et al. (2003) Progressive striatal and cortical dopamine receptor dysfunction in Huntington's disease: a PET study *Brain* 126, 1127–1135.
17. Rosas, H.D., Koroshetz, W.J., Chen, Y.I., Skeuse, C., Vangel, M., Cudkowicz, M.E., et al. (2003) Evidence for more widespread cerebral pathology in early HD: an MRI-based morphometric analysis *Neurology* 60, 1615–1620.
18. Telenius, H., Kremer, H.P., Theilmann, J., Andrew, S.E., Almqvist, E., Anvret, M., et al. (1993) Molecular analysis of juvenile Huntington disease: the major influence on (CAG) n repeat length is the sex of the affected parent *Hum. Mol. Genet.* 2, 1535–1540.
19. Kempermann, G., Kuhn, H.G., and Gage, F.H. (1997) Genetic influence on neurogenesis in the dentate gyrus of adult mice *Proc. Natl Acad. Sci. USA* 94, 10409–10414.

20. Shuttleworth, C.W. and Connor, J.A. (2001) Strain-dependent differences in calcium signaling predict excitotoxicity in murine hippocampal neurons *J. Neurosci.* **21**, 4225–4236.

21. Hockly, E., Cordery, P.M., Woodman, B., Mahal, A., Van, D.A., Blakemore, C., et al. (2002) Environmental enrichment slows disease progression in R6/2 Huntington's disease mice *Ann. Neurol.* **51**, 235–242.

22. MacDonald, M.E., Barnes, G., Srinidhi, J., Duyao, M.P., Ambrose, C.M., Myers, R.H., et al. (1993) Gametic but not somatic instability of CAG repeat length in Huntington's disease *J. Med. Genet.* **30**, 982–986.

23. Rubinsztein, D.C., Leggo, J., Coles, R., Almqvist, E., Biancalana, V., Cassiman, J.J., et al. (1996) Phenotypic characterization of individuals with 30–40 CAG repeats in the Huntington disease (HD) gene reveals HD cases with 36 repeats and apparently normal elderly individuals with 36–39 repeats *Am. J. Hum. Genet.* **59**, 16–22.

24. Snell, R.G., MacMillan, J.C., Cheadle, J.P., Fenton, I., Lazarou, L.P., Davies, P., et al. (1993) Relationship between trinucleotide repeat expansion and phenotypic variation in Huntington's disease *Nat. Genet.* **4**, 393–397.

25. Carter, R.J., Hunt, M.J., and Morton, A.J. (2000) Environmental stimulation increases survival in mice transgenic for exon 1 of the Huntington's disease gene *Mov. Disord.* **15**, 925–937.

26. Kempermann, G., Kuhn, H.G., and Gage, F.H. (1997) More hippocampal neurons in adult mice living in an enriched environment *Nature* **386**, 493–495.

27. Menalled, L.B. (2005) Knock-in mouse models of Huntington's disease *NeuroRx* **2**, 465–470.

28. Hamel, E., Goetz, I.E., and Roberts, E. (1981) Glutamic acid decarboxylase and gamma-aminobutyric acid in Huntington's disease fibroblasts and other cultured cells, determined by a [3H]muscimol radioreceptor assay *J. Neurochem.* **37**, 1032–1038.

29. Saudou, F., Finkbeiner, S., Devys, D., and Greenberg, M.E. (1998) Huntingtin acts in the nucleus to induce apoptosis but death does not correlate with the formation of intranuclear inclusions *Cell* **95**, 55–66.

30. Jackson, G.R., Salecker, I., Dong, X., Yao, X., Arnheim, N., Faber, P.W., et al. (1998) Polyglutamine-expanded human huntingtin transgenes induce degeneration of Drosophila photoreceptor neurons *Neuron* **21**, 633–642.

31. Faber, P.W., Alter, J.R., MacDonald, M.E., and Hart, A.C. (1999) Polyglutamine-mediated dysfunction and apoptotic death of a Caenorhabditis elegans sensory neuron *Proc. Natl Acad. Sci. USA* **96**, 179–184.

32. Karlovich, C.A., John, R.M., Ramirez, L., Stainier, D.Y., and Myers, R.M. (1998) Characterization of the Huntington's disease (HD) gene homologue in the zebrafish Danio rerio *Gene* **217**, 117–125.

33. Miller, V.M., Nelson, R.F., Gouvion, C.M., Williams, A., Rodriguez-Lebron, E., Harper, S.Q., et al. (2005) CHIP suppresses polyglutamine aggregation and toxicity in vitro and in vivo *J. Neurosci.* **25**, 9152–9161.

34. Morton, A.J. and Leavens, W. (2000) Mice transgenic for the human Huntington's disease mutation have reduced sensitivity to kainic acid toxicity *Brain Res. Bull.* **52**, 51–59.

35. Hickey, M.A. and Morton, A.J. (2000) Mice transgenic for the Huntington's disease mutation are resistant to chronic 3-nitropropionic acid-induced striatal toxicity *J. Neurochem.* **75**, 2163–2171.

36. Petersen, A., Chase, K., Puschban, Z., DiFiglia, M., Brundin, P., and Aronin, N. (2002) Maintenance of susceptibility to neurodegeneration following intrastriatal injections of quinolinic acid in a new transgenic mouse model of Huntington's disease *Exp. Neurol.* **175**, 297–300.

37. Zeron, M.M., Hansson, O., Chen, N., Wellington, C.L., Leavitt, B.R., Brundin, P. et al. (2002) Increased sensitivity to N-methyl-D-aspartate receptor-mediated excitotoxicity in a mouse model of Huntington's disease *Neuron* **33**, 849–860.

38. Menalled, L.B. and Chesselet, M.F. (2002) Mouse models of Huntington's disease *Trends Pharmacol. Sci.* **23**, 32–39.

39. Bates, G.P., Mangiarini, L., Mahal, A., and Davies, S.W. (1997) Transgenic models of Huntington's disease *Hum. Mol. Genet.* **6**, 1633–1637.

40. Li, J.Y., Popovic, N., and Brundin, P. (2005) The use of the R6 transgenic mouse models of Huntington's disease in attempts to develop novel therapeutic strategies *NeuroRx* **2**, 447–464.

41. Rubinsztein, D.C. (2002) Lessons from animal models of Huntington's disease *Trends Genet.* **18**, 202–209.

42. Feany, M.B. and Bender, W.W. (2000) A Drosophila model of Parkinson's disease *Nature* **404**, 394–398.

43. Lakso, M., Vartiainen, S., Moilanen, A.M., Sirvio, J., Thomas, J.H., Nass, R. et al. (2003) Dopaminergic neuronal loss and motor deficits in Caenorhabditis elegans overexpressing human alpha-synuclein *J. Neurochem.* **86**, 165–172.

44. Kirik, D. and Bjorklund, A. (2003) Modeling CNS neurodegeneration by overexpression of disease-causing proteins using viral vectors *Trends Neurosci.* **26**, 386–392.

45. Zigmond, M.J. and Stricker, E.M. (1972) Deficits in feeding behavior after intraventricular injection of 6-hydroxydopamine in rats *Science.* **177**, 1211–1214.

46. Barker, R.A. and Dunnett, S.B. (1999) Functional integration of neural grafts in Parkinson's disease *Nat. Neurosci.* **2**, 1047–1048.

47. Sauer, H. and Oertel, W.H. (1994) Progressive degeneration of nigrostriatal dopamine neurons following intrastriatal terminal lesions with 6-hydroxydopamine: a combined retrograde tracing and immunocytochemical study in the rat *Neuroscience* **59**, 401–415.

48. Ungerstedt, U. and Arbuthnott, G.W. (1970) Quantitative recording of rotational behavior in rats after 6-hydroxy-dopamine lesions of the nigrostriatal dopamine system *Brain Res.* **24**, 485–493.

49. Dunnett, S.B., Hernandez, T.D., Summerfield, A., Jones, G.H., and Arbuthnott, G. (1988) Graft-derived recovery from 6-OHDA lesions: specificity of ventral mesencephalic graft tissues *Exp. Brain Res.* **71**, 411–424.

50. Barker, R. and Dunnett, S.B. (1994) Ibotenic acid lesions of the striatum reduce drug-induced rotation in the 6-hydroxydopamine-lesioned rat *Exp. Brain Res.* **101**, 365–374.

51. Iancu, R., Mohapel, P., Brundin, P., and Paul, G. (2005) Behavioral characterization of a unilateral 6-OHDA-lesion model of Parkinson's disease in mice *Behav. Brain Res.* **162**, 1–10.

52. Masliah, E., Rockenstein, E., Veinbergs, I., Mallory, M., Hashimoto, M., Takeda, A., et al. (2000) Dopaminergic loss and inclusion body formation in alpha-synuclein mice: implications for neurodegenerative disorders *Science* **287**, 1265–1269.

53. van der Putten, H., Wiederhold, K.H., Probst, A., Barbieri, S., Mistl, C., Danner, S., et al. (2000) Neuropathology in mice expressing human alpha-synuclein *J. Neurosci.* **20**, 6021–6029.

54. Giasson, B.I., Duda, J.E., Quinn, S.M., Zhang, B., Trojanowski, J.Q., and Lee, V.M. (2002) Neuronal alpha-synucleinopathy with severe movement disorder in mice expressing A53T human alpha-synuclein *Neuron* **34**, 521–533.

55. Lee, M.K., Stirling, W., Xu, Y., Xu, X., Qui, D., Mandir, A.S., et al. (2002) Human alpha-synuclein-harboring familial Parkinson's disease-linked Ala-53 Thr mutation causes neurodegenerative disease with alpha-synuclein aggregation in transgenic mice *Proc. Natl Acad. Sci. USA* **99**, 8968–8973.

56. Fernagut, P.O. and Chesselet, M.F. (2004) Alpha-synuclein and transgenic mouse models *Neurobiol. Dis.* **17**, 123–130.

57. Stein, T.D. and Johnson, J.A. (2002) Lack of neurodegeneration in transgenic mice overexpressing mutant amyloid precursor protein is associated with increased levels of transthyretin and the activation of cell survival pathways *J. Neurosci.* **22**, 7380–7388.

58. Rochet, J.C., Conway, K.A., and Lansbury, P.T., Jr. (2000) Inhibition of fibrillization and accumulation of prefibrillar oligomers in mixtures of human and mouse alpha-synuclein *Biochemistry* **39**, 10619–10626.

59. Dobrossy, M.D. and Dunnett, S.B. (2005) Optimising plasticity: environmental and training associated factors in transplant-mediated brain repair *Rev. Neurosci.* **16**, 1–21.

60. Watts, C., Caldwell, M.A., and Dunnett, S.B. (1998) The development of intracerebral cell-suspension implants is influenced by the grafting medium *Cell Transplant.* **7**, 573–583.

61. Johann, V., Schiefer, J., Sass, C., Mey, J., Brook, G., Kruttgen, A., et al. (2007) Time of transplantation and cell preparation determine neural stem cell survival in a mouse model of Huntington's disease *Exp. Brain Res.* **177**, 458–470.

62. Watts, C., McNamara, I.R., and Dunnett, S.B. (2000) Volume and differentiation of striatal grafts in rats: relationship to the number of cells implanted *Cell Transplant.* **9**, 65–72.

63. O' Keeffe, G.W. and Sullivan, A.M. (2005) Donor age affects differentiation of rat ventral mesencephalic stem cells *Neurosci. Lett.* **375**, 101–106.

64. Scherzinger, E., Lurz, R., Turmaine, M., Mangiarini, L., Hollenbach, B., Hasenbank, R., et al. (1997) Huntingtin-encoded polyglutamine expansions form amyloid-like protein aggregates in vitro and in vivo *Cell* **90**, 549–558.

65. Harvey, A.R., Symons, N.A., Pollett, M.A., Brooker, G.J., and Bartlett, P.F. (1997) Fate of adult neural precursors grafted to adult cortex monitored with a Y-chromosome marker *Neuroreport* **8**, 3939–3943.

66. Miller, M.W. and Nowakowski, R.S. (1988) Use of bromodeoxyuridine-immunohistochemistry to examine the proliferation, migration and time of origin of cells in the central nervous system *Brain Res.* **457**, 44–52.

67. Caldwell, M.A., He, X., and Svendsen, C.N. (2005) 5-Bromo-2'-deoxyuridine is selectively toxic to neuronal precursors in vitro *Eur. J. Neurosci.* **22**, 2965–2970.

68. Christie, B.R. and Cameron, H.A. (2006) Neurogenesis in the adult hippocampus *Hippocampus* **16**, 199–207.

69. Bull, N.D. and Bartlett, P.F. (2005) The adult mouse hippocampal progenitor is neurogenic but not a stem cell *J. Neurosci.* **25**, 10815–10821.

70. Burns, T.C., Ortiz-Gonzalez, X.R., Gutierrez-Perez, M., Keene, C.D., Sharda, R., Demorest, Z.L., et al. (2006) Thymidine analogs are transferred from prelabeled donor to host cells in the central nervous system after transplantation: a word of caution *Stem Cells* **24**, 1121–1127.

71. Hendriks, W.T., Ruitenberg, M.J., Blits, B., Boer, G.J., and Verhaagen, J. (2004) Viral vector-mediated gene transfer of neurotrophins to promote regeneration of the injured spinal cord *Prog. Brain Res.* **146**, 451–476.

72. Follenzi, A. and Naldini, L. (2002) HIV-based vectors. Preparation and use *Methods Mol. Med.* **69**, 259–274.

73. Lewis, P.F. and Emerman, M. (1994) Passage through mitosis is required for oncoretroviruses but not for the human immunodeficiency virus *J. Virol.* **68**, 510–516.

74. Liu, Y., Himes, B.T., Solowska, J., Moul, J., Chow, S.Y., Park, K.I., et al. (1999) Intraspinal delivery of neurotrophin-3 using neural stem cells genetically modified by recombinant retrovirus *Exp. Neurol.* **158**, 9–26.

75. Limke, T.L. and Rao, M.S. (2002) Neural stem cells in aging and disease *J. Cell Mol. Med.* **6**, 475–496.

76. Watts, C. and Dunnett, S.B. (1998) Effects of severity of host striatal damage on the morphological development of intrastriatal transplants in a rodent model of Huntington's disease: implications for timing of surgical intervention *J. Neurosurg.* **89**, 267–274.

77. Svendsen, C.N., ter Borg, M.G., Armstrong, R.J., Rosser, A.E., Chandran, S., Ostenfeld, T., et al. (1998) A new method for the rapid and long term growth of human neural precursor cells *J. Neurosci. Methods* **85**, 141–152.

Chapter 11

Animal Models of Multiple Sclerosis

Roberto Furlan, Carmela Cuomo, and Gianvito Martino

Summary

Since its first description, experimental autoimmune encephalomyelitis, originally designated experimental allergic encephalitis (EAE), has been proposed as animal model to investigate pathogenetic hypotheses and test new treatments in the field of central nervous system inflammation and demyelination, which has become, in the last 30 years, the most popular animal model of multiple sclerosis (MS). This experimental disease can be obtained in all mammals tested so far, including nonhuman primates, allowing very advanced preclinical studies. Its appropriate use has led to the development of the most recent treatments approved for MS, also demonstrating its predictive value when properly handled. Some of the most exciting experiments validating the use of neural precursor cells (NPCs) as a potential therapeutic option in CNS inflammation have been performed in this model. We review here the most relevant immunological features of EAE in the different animal species and strains, and describe detailed protocols to obtain the three most common clinical courses of EAE in mice, with the hope to provide both cultural and practical basis for the use of this fascinating animal model.

Key words: Experimental autoimmune encephalomyelitis, Multiple sclerosis, Disease induction.

1. Introduction

Experimental autoimmune encephalomyelitis (EAE) is possibly the best animal model to study autoimmune diseases and in particular demyelinating diseases of the central nervous system (CNS) such as multiple sclerosis (MS). Since the first classical studies of Rivers *(1)*, in monkeys immunized with CNS homogenate, EAE has been an invaluable tool for dissecting mechanisms of the immune response against self-antigens within the CNS, as well as to test new therapies for the treatment of autoimmune diseases. An autoimmune response leading to EAE in susceptible

Neil J. Scolding and David Gordon (eds.), *Methods in Molecular Biology, Neural Cell Transplantation, vol. 549*
DOI: 10.1007/978-1-60327-931-4_11, © Humana Press, a part of Springer Science+Business Media, LLC 2009

species can be obtained by means of active immunization with CNS proteins and by passive transfer of T lymphocytes reacting against myelin antigens to syngeneic recipients. The role of T lymphocytes in EAE was first demonstrated by Paterson who succeeded in transferring disease by means of T cells from immunized animals (2). Since then, many researchers have attempted to characterize the role of T cells in EAE. Over the years, it became clear that activated CD4+ T cells mediate EAE upon recognition of the target antigen bound to class II molecules of the major histocompatibility complex (MHC) (3). Encephalitogenic T cells can be retrieved from the blood of immunized as well as naive animals, supporting the concept that auto-aggressive lymphocytes are part of the natural immune repertoire (4, 5).

Several species and strains have been utilized including mice, rats, and guinea pigs. Virtually all mammalian species are susceptible to EAE as long as they are properly immunized. The clinical, pathological, and immunological picture of autoimmune models of demyelination depends upon the mode of sensitization, the nature of the immunogen, and the genetic background of each species and strain.

Modes of sensitization include the route of immunization, primarily subcutaneously, and the use of immunogens emulsified with an equal volume of complete Freund's adjuvant containing *Mycobacterium tubercolosis* to create an antigen depot. Boosts with *Bordetella pertussis* are often used to help in opening the blood–brain barrier (BBB).

Whole myelin homogenate as well as distinct myelin proteins, including myelin basic protein (MBP), myelin oligodendrocyte glycoprotein (MOG), and proteolipid protein (PLP), have been used to induce EAE in different species. The role of MBP, an abundant hydrophilic myelin protein, was first characterized by Ben-Nun and colleagues, who successfully induced EAE by transferring MBP-specific T-cell lines to naive Lewis rats (6). Passive transfer studies also elucidated the role of other myelin antigens, including PLP (7) and MOG (8). MOG is a minor glycoprotein exposed on the surface of the myelin sheath and is the target of both humoral and cellular immune responses. The importance of the humoral response to MOG was first demonstrated by Linington who showed the presence of extensive demyelination following the injection of anti-MOG auto-antibodies into animals with T-cell-mediated EAE (9). Other encephalitogenic proteins include a lipidated form of MBP (10) as well as nonmyelin autoantigens such as the astrocyte-derived calcium binding protein S100β (11).

Depending upon the species, the antigen and the mode of sensitization, EAE may have a monophasic acute course (**Fig. 1a**), as well as a chronic relapsing (**Fig. 1b**) and even a primary progressive (**Fig. 1c**) course that mimics human MS. The classical

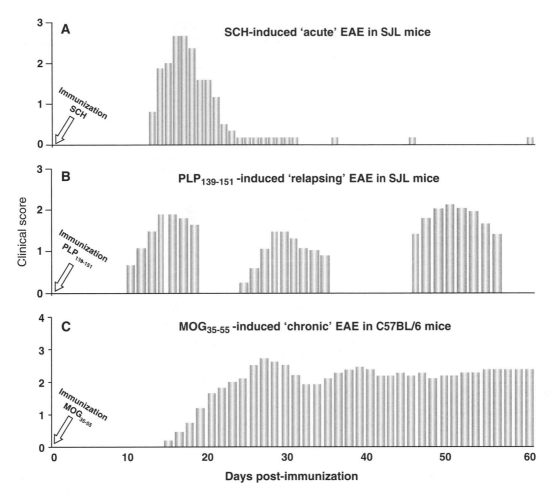

Fig. 1. EAE clinical courses.

picture of acute EAE is characterized by perivascular inflammation mainly represented by CD4+ and CD8+ T lymphocytes and macrophages within the spinal cord and less consistently in the brain. Nevertheless, the manipulation of the above-mentioned factors can lead to a wide spectrum of neuropathological patterns inclusive of demyelination, remyelination, gliosis, loss of axons, and in certain species also necrosis *(12)*.

1.1. EAE in Inbred Species

Olitsky was the first to establish EAE in mice *(13)* and since then, thousands of scientists have used inbred rodents as the most suitable species for EAE studies. EAE has been successfully induced in guinea pigs *(14)* as well as in several strains of rats and mice. Some mice strains such as SJL/J develop a relapsing–remitting disease following active immunization with spinal cord homogenate

(SCH) *(15)*. Pathology shows the presence of mononuclear cell infiltrates within the CNS and demyelination. A similar disease can also be obtained by the adoptive transfer of MBP-specific T lymphocytes *(16)*. In this H-2^S strain, encephalitogenic T cells recognize few epitopes of MBP concentrated within the sequence 89–101, preferentially using the T-cell receptor (TCR) Vβ17 segment *(17)*. A more restricted response to encephalitogenic determinants of MBP has been reported for H-2^U mice susceptible to EAE such as PL/J and B.10.PL. Both encephalitogenic and non-encephalitogenic T cells recognize the N-terminal peptide Ac1-9 utilizing the same TCR Vβ8.2/Vα2 (or Vα4) gene combination *(3)*. Interestingly, the same TCR gene segments are characteristic of the T-cell response to MBP in Lewis rats *(18)*. In Lewis rats, EAE induced either by active immunization or by passive transfer of T cells is an acute monophasic disease mostly characterized by inflammation *(19)*. In this species, EAE is mediated by CD4$^+$ encephalitogenic T lymphocytes specific for the peptide 68–88, in the context of the class II molecule RT1.B^1 *(20)*, which rapidly home and persist at the site of inflammation *(21)*. The dominant use of a restricted TCR repertoire has been successfully exploited in designing immunospecific therapies by means of anti-Vβ monoclonal antibodies *(22)* and active vaccination with epitopes of the encephalitogenic TCR molecule *(23)*. Nevertheless, the presence of an epitope dominance, as well as limited TCR usage within the T-cell response to MBP, seems to be confined largely to the early phases of EAE and remains controversial in humans. Later stages of EAE are characterized by a more diverse recognition of previously cryptic determinants inside the MBP molecule (intramolecular spreading) and within other CNS antigens (intermolecular spreading) *(24)*. The appearance of a diverse T-cell repertoire is further confirmed by the presence, in the spinal cord of rats with EAE, of a heterogeneous Vβ population during the recovery phase of disease *(25)*. This may result from apoptosis of encephalitogenic T-cell clones *(26)* followed by secondary recruitment of activated cells from the recirculating T-cell pool specific for minor antigens.

Recently, a chronic relapsing form of EAE has been successfully induced in Biozzi (AB/H) mice by immunization with SCH *(27)* or MOG *(28)*. In this species, pathology shows the full spectrum of lesions seen in MS including sharp demyelination and remyelination. Another useful model was recently described in DA rats which are characterized by a chronic relapsing course, inflammation, and demyelination *(29)*.

In contrast to the rather restricted response to MBP, a diverse recognition of determinants within other myelin antigens, such as PLP *(30)* and MOG *(28)*, seems to occur in most rodents. Although as a general rule for all inbred species it appears that strains of different haplotype react with different epitopes of myelin antigens.

The presence of encephalitogenic T cells reacting against self-antigens in naive animals is of remarkable conceptual importance and supports the fact that autoaggressive T cells escaping thymic deletion are maintained within the normal circulating T-lymphocyte pool *(4, 5)*. A possible explanation for self-reactive T cells escaping negative selection has been recently clarified in an MBP–/– transgenic model in which endogenous MBP inactivated high avidity clones reacting against the immunodominant epitope making that determinant appear cryptic *(31)*.

Thus, other factors are also necessary to cause an autoimmune disease. The genetic background is a major factor conferring susceptibility to EAE and a number of murine loci have already been identified *(32)*. Both MHC and non-MHC genes have been reported to control the development and severity of EAE *(33)*. The role of environmental factors has been elegantly elucidated by Goverman and colleagues, who created transgenic mice expressing an MBP-specific TCR which spontaneously developed EAE only when challenged with microbial stimuli *(34)*. In a similar model, the complete cohort of anti-MBP TCR transgenic mice, deficient for mature T and B cells, developed spontaneous EAE, suggesting that other cells may have a protective role counteracting with encephalitogenic cells *(35)*. A protective and/or regulatory role has been claimed for almost all cells involved in the immune response including CD4$^+$ T cells *(36)*, CD8+ *(37)*, CD4– CD8– *(38)*, macrophages *(39)*, B cells *(40)*, γδ T cells *(41)*, and NK cells *(42)*.

The environment exerts a major effect on EAE, leading to the activation of potentially autoaggressive T cells which consequently home to the brain and induce disease *(43)*. Activation state is the necessary prerequisite for T cells to migrate through the BBB irrespective of their antigen specificity. Activation of myelin-specific T lymphocytes in the peripheral compartment can occur through a mechanism of molecular mimicry *(44)* and by stimulation with microbial superantigens *(45)*. Other theoretical possibilities such as activation of T cells carrying two sets of receptors, one specific for a foreign protein and another for a self-antigen, has never been demonstrated to play a role in autoimmunity *(46)*. Migration through the BBB requires the involvement of adhesion molecules on both T cells (LFA-1 and VLA-4) and endothelial cells (ICAM-1 and VCAM-1) *(47)*. Following migration through the BBB, neuroantigen-specific CD4$^+$ T cells are reactivated in situ by fragments of myelin antigens presented in the framework of MHC class II molecules on the surface of local antigen-presenting cells including macrophages, microglia, and less efficiently also by astrocytes *(48)*. These events are associated with the release of proinflammatory cytokines leading to the up-regulation of MHC molecules on a variety of resident, antigen-presenting cells *(49)*. The kinetics of cytokines along the EAE clinical course suggest the role of

T-helper 1 (Th1) cytokines such as TNF-β, IL-12, TNF-α, and IFN-γ before and at the peak of disease. The recovery phase correlates with Th2 cytokines such as TGF-β, IL-10 and possibly IL-4 *(50)*. The onset of overt inflammation maintains also endothelial activation and leads to a second wave of inflammatory recruitment, including T cells, macrophages that result in tissue demage by means of TNF-α *(51)*, oxygen and nitrogen intermediates, perforin, and complement *(52)* and demyelinating antibodies *(9)*.

Thus far, no consistent evidence differentiates between activated MBP-specific T lymphocytes with no encephalitogenic capabilities and their pathogenic counterparts, in spite of identical growing and stimulation conditions, sharing of epitope specificity and MHC restriction. Some studies suggest that differences in encephalitogenicity correlate with a predominant Th1 cytokine profile *(53)*, their brain homing capacity *(54)*, and the ability to mediate a delayed type hypersensitivity (DTH) response *(55)*. Moreover, encephalitogenic T cells, despite their CD4+ phenotype, have been demonstrated to be cytotoxic for cells, such as astrocytes, presenting myelin antigens in an MHC-restricted manner *(56)*. On the other hand, it has been suggested that nonencephalitogenic T cells can initiate autoimmune regulatory mechanisms through the production of IL-3 *(57)*. It has been postulated that, at least in Lewis rats, encephalitogenicity of MBP-specific T cells correlates with the cytokine profile which depends on the MHC haplotype of the strain *(58)*.

A rather simplistic picture of the EAE cytokine network would suggest that TNF-α, TNF-β, IFN-γ, and IL-12 (proinflammatory cytokines) have a disease-promoting role, while TGF-β, IL-10, and possibly IL-4 (anti-inflammatory cytokines) may protect from disease. Although a detailed analysis of the current literature on this topic is far beyond the scope of this chapter, an enormous amount of date states that the real picture is much more complicated. Several factors influence the cytokine profile of effector and regulatory cells in EAE and, therefore, the final outcome of the immune response within the target organ. These include the age of the animal *(59)*, nature of antigen-presenting cells *(60, 61)*, local cytokine micro-environment *(62)*, selective engagement with co-stimulatory molecules *(63)*, interaction with altered forms of the immunizing antigen *(64)*, and the route of immunization *(65)*.

Based on the hypothesis that Th1 cytokines play a promoting effect on autoimmunity while Th2 cytokines may have a protective role, immune deviation toward a Th2 profile has been exploited for successfully treating EAE by administration of anti-inflammatory cytokines *(66)*, altered peptide ligands *(64)*, monoclonal antibodies (MAb) affecting B7/CD28 interactions *(63)*, anti-inflammatory cytokines *(67, 68)*, and by induction of oral tolerance *(65)*.

Despite the success of most of the experimental treatments targeting Th1 cytokines, the Th1 vs. Th2 dichotomy underscores the complexity of interactions that lead to reciprocal cross-regulation of Th1/Th2 responses. This has been dramatically elucidated by the severe aggravation of primates EAE due to an enhanced Th2 response occurring after discontinuation of treatment for immune deviation *(69)*. Moreover, conflicting results arise from the utilization of genetically manipulated mice either lacking or over-expressing cytokines, as is the case in a number of TNF-α studies which demonstrated a different clinical phenotype depending on the experimental conditions. For example, over-expression of TNF-α in transgenic mice leads to spontaneous inflammation and demyelination within the CNS *(70)*, while TNF-α-deficient knock-out mice can still develop EAE, thus challenging the role of this cytokine in EAE pathogenesis *(71)*. Surprisingly, in a recent study, MOG-immunized, TNF-deficient mice developed a severe EAE but were remarkably ameliorated by the administration of TNF, also possibly supporting a protective role for this cytokine *(72)*. Targeting cytokine genes has helped to elucidate the role of other cytokines, such as INF-γ *(73)*, IL-4 and IL-10 *(74)*, nitric oxide *(75)*, and also on Fas/Fas-ligand and perforin pathways *(76)*. Nevertheless, it must be kept in mind that results of genetic manipulation result in experimental conditions that only partially represent the in vivo situation and that apparently underscore the redundancy of the cytokine system.

Overall, the deep knowledge of the immunogenetics of inbred species, the possibility of successfully manipulating it, together with their accessibility and costs, still make these species the first choice for studies on autoimmune diseases of the CNS.

1.2. EAE in Outbred Species

EAE in nonhuman primates represented the first experimental model for demyelinating diseases of the CNS *(1)*. The recent advances in housing and handling techniques, knowledge of primate anatomy, immunology, and genetics and the compatibility of most of human reagents and diagnostic techniques have sparked wide interest on EAE in these species. A unique advantage of monkeys arises from their outbred condition that closely resembles the human status. Moreover, the transfer of immunocompetent cells in outbred primates is allowed by the possibility of crossing the trans-species barrier among closely related species *(77)* and by the natural bone marrow chimerism in some others *(5)*. Therefore, in primates it is possible to elucidate the role of pathogenic cells by means of passive transfer experiments in a polymorphic setting. Recently, EAE has been induced in the common marmoset *Callithrix jacchus*, a unique primate species that develops in utero as genetically distinct twins or triplets sharing bone marrow-derived elements through a common placental circulation *(78)*. It has been recently demonstrated that *C. jacchus* TCR genes are extensively conserved *(79)* and that MHC class II region genes, despite a relatively low polymorphism, encode the evolutionary equivalents of the HLA-DR and -DQ molecules *(80)*. A fully demyelinating form of EAE has been induced either by active immunization with whole myelin *(78)*, myelin oligodendrocyte glycoprotein (MOG) or by MBP followed by administration of MOG-specific antibodies *(81)*. Passive transfer experiments have demonstrated that encephalitogenic MBP-specific T cells are part of the normal marmoset repertoire *(5)*. Similarly to humans, MBP-reactive T cells recognize different determinants by means of a diverse TCR repertoire (Antonio Uccelli, personal communication). Pathology of EAE induced with whole myelin or MOG is characterized by perivascular inflammation with conspicuous primary demyelination whose topography correlates with MRI abnormalities *(82)*. On the contrary, immunization with MBP or passive transfer of MBP-reactive T cells leads to mild inflammation and no demyelination (*5*, Mancardi and Uccelli unpublished results). A complex role of cytokines, possibly released by activated T lymphocytes and macrophages during the immune response, is suggested by the high expression of CD40 and CD40-ligands in marmoset active lesions *(83)*. The role of Th1 cytokines has been demonstrated by the prevention of disease following treatment with the cAMP-specific type IV phosphodiesterase inhibitor Rolipram *(84)*. On the other hand, the ambiguous role of Th2 cytokines was highlighted by the enhancement of EAE occurring after discontinuation of a MOG-based tolerization treatment due to an enhanced proliferative and antibodies response to the antigen. Hence, it is likely that Th2-like T cells play a different role, protecting or favoring autoimmunity, under different conditions *(69)*.

EAE has also been induced in macaques by immunization with myelin or MBP emulsified in complete adjuvant *(85)*. The EAE course is primarily hyperacute or acute, often with a lethal outcome, and is characterized by intense inflammation associated with hemorrhages and necrosis resembling acute disseminated encephalomyelitis *(86)*. The association of EAE susceptibility with an MHC class II allele *(87)*, the presence of myelin-reactive T cells correlating with the course of EAE *(86)*, the beneficial effect of anti-CD4 antibodies on EAE outcome *(88)* and the possibility of inducing a mild form of EAE by adoptive transfer of MBP-specific T cells from unprimed animals *(89)* all provide strong evidence that T cells play a central role in this model. At the moment, the major advantage of a nonhuman primate EAE model for human MS resides in the molecular and functional organization of the primate immune system leading to the possibility of evaluating the safety and efficacy of biological molecules as therapy for MS.

1.3. Conclusions	EAE induced in mice and nonhuman primates represents a valuable tool to investigate novel therapeutic approaches in human MS, characterized by neural tissue damage occurring in the context of a very complex immune-mediated inflammatory reaction. In the following sections, we will describe the protocols to obtain (a) acute monophasic EAE **(Fig. 1a)** induced in SJL mice by immunization with SCH; (b) relapsing–remitting EAE **(Fig. 1b)** induced in SJL mice by immunization with PLP139–151; and (c) chronic-progressive EAE **(Fig. 1c)** induced in C57BL/6 mice by immunization with MOG35–55.

2. Materials

2.1. SCH Preparation	1. 10-ml Glass syringes.
	2. 21-Gauge needles.
	3. Surgical instruments for mouse dissection.
	4. 4. 70-μm Falcon cell strainer.
	5. Lyophilizer.
2.2. EAE Induction	1. 10-ml Glass syringes.
	2. 30-ml Glass tubes (corex).
	3. 21-Gauge needles.
	4. Incomplete Freund's Adjuvant (IFA) (Sigma).
	5. Heat-inactivated, lyophilized *Mycobacterium tuberculosis* strain H37R (Beckton Dickinson, formerly DIFCO).
	6. 1-ml Syringes with removable 25-gauge needle.
	7. MOG, residues 35–55 (MEVGWYRSPFSRVVHLYRNGK), or proteolipid proptein (PLP) residues 139–151 (HSLGKALGHP-DKF) purity over 95% (*see* **Note 1**), or SCH (see above).
	8. Pertussis toxin (pancreatic islet activating) from *Bordetella pertussis* (Sigma).
	9. Female C57BL/6 or SJL mice (depending on the antigen used), 18–20 g (6–8-weeks old).

3. Methods

3.1. SCH Preparation	1. SCH is prepared according to Levine and Sowinski *(90)*. Using surgical scissors, take out cervical, thoracic, and lumbar column from sacrificed mice.

2. Firmly mount the 21-gauge needle on the 10-ml syringe filled with saline and flush out the spinal cords from the excised vertebral columns.

3. Homogenize the tissue by straining through the 70-μm Falcon cell strainer using a syringe piston.

4. Weigh the crude SCH and add saline 4:1 (w/v; es. 1,000 mg spinal cords with 250 μl saline).

5. Lyophilize according to the standard protocol of the available lyophilizer.

6. Store at –20°C.

3.2. EAE Induction

1. *Emulsion preparation.* For the emulsion's final volume, calculate 300 μl/mouse and at least 2–3 ml excess. In a glass tube, re-suspend the lyophilized antigen (MOG35–55, PLP139–151, or SCH) in 1× PBS at concentration indicated in **Table 1** (*see* **Note 2**).

2. In a second glass tube, prepare a homogeneous suspension of 8 mg/ml *M. tuberculosis* in IFA to obtain CFA. The lyophilized *M. tuberculosis* does not dissolve in IFA. An homogenous suspension can be obtained by careful stirring with a glass syringe, but *M. tuberculosis* will immediately re-precipitate at the bottom of the tube (*see* **Note 3**).

3. In a third glass tube, mix the antigen, resuspended in PBS, and the CFA in a 1:1 volume ratio. Mix the emulsion by aspirating/expelling with a glass syringe through a 21-gauge needle until it becomes firm, dense, and very creamy (*see* **Note 4**). This usually takes at least 5–10 min mixing.

4. When the emulsion is ready, fill the 1-ml syringes by aspirating the emulsion through the 21-gauge needle you used to mix it, remove it, place back the original 25-gauge needle (*see* **Note 5**). Then, swing the syringe in order to have the emulsion towards the piston and push the latter until the dead volume of the syringe is filled. If properly done, this will take you to have 900 μl emulsion in the syringe, exactly the amount needed for three mice.

5. Inject mice with the emulsion according to **Table 1** (*see* **Notes 6** and **7**).

6. Inject each mouse intravenously with P.T. re-suspended in 100 μl of PBS (*see* **Note 8**), according to the dose and schedule indicated in **Table 1**.

3.3. EAE Clinical Evaluation

1. Clinical score and body weight is recorded daily.

2. Mice will have usually 1–2 g of weight loss in the days preceding clinical onset, which usually occurs, variably, between the 9th and 18th day after immunization.

Table 1
Antigen concentrations and immunization schedules

EAE type	Mouse strain	Antigen used	Antigen concentration	Sites of injection	P.T.	Boost
Acute monophasic	SJL	SCH	27 mg/ml (4 mg/mouse)	The two hind footpads (approx. 50-μl each) and under the skin of the neck (remaining 200 μl)	400 ng/mouse the day of immunization and 48 h later	No
Relapsing–remitting	SJL	PLP139–151	2 mg/ml (300 μg/mouse)	Under the skin of one flank	500 ng/mouse the day of immunization and 48 h later	After seven days, injecting the emulsion on the opposite flank and repeating P.T. the day of the boost and 48 h later
Chronic	C57BL/6	MOG35-55	1.3 mg/ml (200 μg/mouse)	Under the skin of both flanks and at the tail base, 100 μl at each site	500 ng/mouse the day of immunization and 48 h later	No

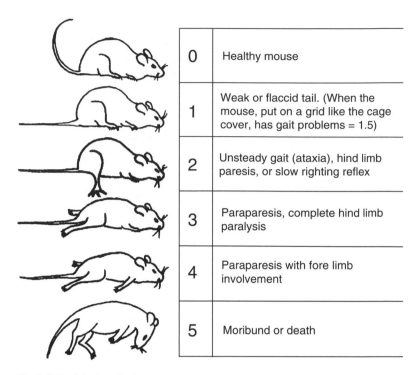

0	Healthy mouse
1	Weak or flaccid tail. (When the mouse, put on a grid like the cage cover, has gait problems = 1.5)
2	Unsteady gait (ataxia), hind limb paresis, or slow righting reflex
3	Paraparesis, complete hind limb paralysis
4	Paraparesis with fore limb involvement
5	Moribund or death

Fig. 2. EAE clinical evaluation.

3. A nonparametric scale with five steps is used for clinical evaluation (**Fig. 2**). Mice are taken from the cage by the tail and observed from the abdominal side: healthy mice extend their hind limbs very wide when held in this position. An asymmetric position, especially with one of the hind limbs retracted close to the abdomen, may represent an initial sign of disease. A weak tail and an unsteady gait are the signs at onset. To reveal an uncertain gait, it is useful to observe mice walking on a grid, like the cage cover. Healthy mice are quick in movements on a grid almost as on firm ground. Sick mice are unsure, move slowly, and often stumble. In the absence of other signs, we score the uncertain gait on the cage cover as 1.5. Another useful examination is that of the righting reflex. Trying to put healthy mice on its back results in such a quick flipping to the upright position that it is impossible to see the abdomen. A sick mice may have a slow, very slow, to complete impaired righting reflex. In the absence of other signs, we score any grade of impairment of the righting reflex as 2. Paraparesis (score 3) and involvement of fore limbs (score 4) are easily identified (*see* **Note 9**).

4. Notes

1. Quality of the peptide is the most relevant issue to obtain a good disease. Incomplete incidence or very low severity of EAE are almost invariably related to peptide's quality.

2. In order to avoid waste of material due to weighing procedures, we usually dissolve the lyophilized MOG35–55 or PLP139–151 peptide in 1× PBS at 2 mg/ml, sterilize by filtration on a 0.2-μm syringe filter (for in vitro immunological assays), and store it at −80°C in 1-ml aliquots for several months, up to 1 year, without loss of encephalitogenicity.

3. To improve the suspension, *M. tuberculosis* can be further triturated in a mortar with a pestel to obtain a more subtle powder, before mixing to IFA.

4. To check the emulsion's quality, let a drop of it fall in water: a good emulsion stays compact without dissolving and without blurred edges.

5. Since the emulsion is very dense, a great pressure has to be applied to push the needed amount out of the syringe. This can lead to the popping off of the needle, and sometimes the emulsion comes out with great pressure and can get in the operator's eyes. Therefore, great care has to be taken to put the 25-gauge needle back to the 1-ml syringe pressing it firmly. Additionally, the connection can be stabilized with a small piece of parafilm.

6. Mice should be housed in an SPF animal facility or, at least, in a clean conventional room since viral, bacterial, and especially parasitic infections can alter the disease course (Roberto Furlan personal observation).

7. Injections under the skin of the flanks is performed at the thoracic level to avoid the possibility to inject the emulsion intraperitoneally: consequent peritoneal inflammation leads to altered gait which is often confused with the clinical manifestation of EAE.

8. Pertussis toxin can be stored at −80°C in aliquots at the final concentration needed, for at least 6 months, with only a minor loss of activity.

9. When we have mice with grade 3 or 4 EAE, we put wet food pellets in the cage to allow paraplegic mice to eat and survive.

References

1. Rivers TM, Sprunt DH, Berry GP (1933) Observations on the attempts to produce acutedisseminated encephalomyelitis in monkeys. J Exp Med 58:39–53

2. Paterson PY (1960) Transfer of allergic encephalomyelitis in rats by means of lymph node cells. J Exp Med 111:119–135

3. Zamvil SS and Steinman L (1990) The T lymphocyte in experimental allergic encephalomyelitis. Annu Rev Immunol 8:579–621

4. Schlusener HJ and Wekerle H (1985) Autoaggressive T-lymphocyte lines recognizing the encephalitogenic portion of myelin basic protein: in vitro selection from unprimed rat T-lymphocyte populations. J Immunol 135:3128–3133

5. Genain CP, Lee-Parritz D, Nguyen MH, Massacesi L, Joshi N, Ferrante R, Hoffman K, Moseley M, Letvin NL, Hauser SL (1994) In healthy primate, circulating autoreactive T cells mediate autoimmune disease. J Clin Invest 94:1339–1345

6. Ben-Nun A, Wekerle H, Cohen IR (1981) The rapid isolation of clonable antigen specific T lymphocyte linescapable of mediating auto immune encephalomyelitis. Eur J Immunol 11:195–199

7. Yamamura T, Namikawa T, Endoh M, Kunishita T, Tabira T (1986) Passive transfer of experimental allergic encephalomyelitis induced by proteolipid apoprotein. J Neurol Sci 76:269–275

8. Linington C, Berger T, Perry L, Weerth S, Hinze-Selch D, Zhang Y, Lu HC, Lassmann H, Wekerle H (1993) T cell specific for the myelin oligodendrocyte glycoprotein (MOG) mediate an unusual autoimmune inflammatory response in the central nervous system. Eur J Immunol 23:1364–1372

9. Linington C, Bradl M, Lassmann H, Brunner C, Vass K (1988) Augmentation of demyelination in rat acute allergic encephalomyelitis by circulating mouse monoclonal antibodies directed against a myelin/oligodendrocyte glycoprotein. Am J Pathol 130:443–454

10. Massacesi L, Vergelli M, Zehetbauer B, Liuzzi GM, Olivotto J, Ballerini C, Uccelli A, Mancardi GL, Riccio P, Amaducci L (1993) Induction of experimental autoimmune encephalomyelitis in rats and immune response to myelin basic protein in lipid bound form. J Neurol Sci 119:91–98

11. Kojima K, Berger T, Lassmann H, Hinze-Selch D, Zhang Y, Gehrmann J, Reske K, Wekerle H, Linington C (1994) Experimental autoimmune panencephalitis and uveoretinitis transferred to the Lewis rat by T lymphocytes specific for the S100β molecule, a calcium binding protein of astroglia. J Exp Med 180:817–829

12. Raine CS (1997) The lesion in multiple sclerosis and chronic relapsing experimental allergic encephalomyelitis: a structural comparison. In Raine CS, McFarland HF, Tourtellotte WW (eds) Multiple sclerosis: clinical and pathogenetic basis. Chapman and Hall, London, pp 242–286

13. Olitsky PK and Yager RH (1949) Experimental disseminated encephalomyelitis in mice. J Exp Med 90:213–223

14. Lisak RP, Zweiman B, Kies MW, Driscoll B (1975) Experimental allergic encephalomyelitis in resistant and susceptible guinea pigs: in vivo and in vitro correlates. J Immunol 114:546–549

15. Brown AM and McFarlin DE (1981) Relapsing experimental allergic encephalomyelitis in the SJL/J mouse. Lab Invest 45:278–284

16. Mokhtarian F, McFarlin DE, Raine CS (1984) Adoptive transfer of myelin basic proteinsensitized T-cells produces chronic relapsingdemyelinating disease in mice. Nature 309:356–358

17. Sakai K, Sinha AA, Mitchell DJ, Zamvil SS, Rothbard JB, McDevitt HO (1988) Involvement of distinct murine T-cell receptors in the autoimmune encephalitogenic response to nested epitopes of myelin basic protein. Proc Natl Acad Sci USA 85:8608–8612

18. Burns FR, Li RX, Shen L, Offner H, Chou K, Vandenbark A, Haber-Katz E (1989) Both rats and mouse T-cell receptors specific for the encephalitogenic determinant of myelin basic protein use similar Vα and Vβ chain genes even though the major hystocompatibility complex and encephalitogenic determinants being recognized are different. J Exp Med 169:27–39

19. Lassmann H, Brunner C, Bradl M, Linington C (1988) Experimental allergic encephalomyelitis: the balance between encephalitogenic T lymphocytes and demyelinating antibodies determines size and structure of demyelinated lesions. Acta Neuropathol 75:566–576

20. Vandenbark AA, Offner H, Reshef T, Fritz RB, Chou C, Cohen IR (1985) Specificity of T lymphocyte lines for peptides of myelin basic protein. J Immunol 139:229–233

21. Kim G, Tanuma N, Kojima T, Kohyama K, Suzuki Y, Kawazoe Y, Matsumoto Y (1998) CDR3 size spectratyping and sequencing of spectratype-derived TCR of spinal cord T cells in autoimmune encephalomyelitis. J Immunol 160:509–513

22. Zaller DM, Osman G, Kanagawa O, Hood L (1990) Prevention and treatment of murine experimental allergic encephalomyelitis with

T-cell receptor Vβ specific antibodies. *J Exp Med* **171**:1943–1955

23. Howell MD, Winters ST, Olee T, Powell HC, Carlo DJ, Brostoff SW (1989) Vaccination against experimental encephalomyelitis with T cell receptor peptides. *Science* **246**:668–670

24. Lehmann PV, Forsthuber T, Miler A, Sercarz EE (1992) Spreading of T cell autoimmunity to cryptic determinants of an autoantigen. *Nature* **358**:155–157

25. Offner H, Buenafe AC, Vainiene M, Celnik B, Weinberg AD, Gold DP, Hashim G, Vandenbark AA (1993) Where, when and how to detect biased expression of disease relevant Vβ genes in rats with experimental autoimmune encephalomyelitis. *J Immunol* **151**:506–517

26. Bauer J, Wekerle H, Lassmann H (1995) Apoptosis in brain-specific autoimmune disease. *Curr Opinion Immunol* **7**:839–843

27. Baker D, O'Neill JK, Gschmeissner SE, Wilcox CE, Butter C, Turk JL (1990) Induction of chronic relapsing experimental allergic encephalomyelitis in Biozzi mice. *J Nueroimmunol* **28**:261–270

28. Amor S, Groome N, Linington C, Morris MM, Dornmair K, Gardinier MV, Matthieu JM, Baker D (1994) Identification of epitopes of myelin oligodendrocyte glycoprotein for the induction of experimental allergic encephalomyelitis in SJL and Biozzi AB/H mice. *J Immunol* **153**:4349–4356

29. Lorentzen JC, Issazadeh S, Storch M, Mustafa MI, Lassman H, Linington C, Klareskog L, Olsson T (1995) Protracted relapsing and demyelinating experimental autoimmune encephalo myelitis in DA tats immunized with syngeneic spinal cord and incomplete Freund's adjuvant. *J Neuroimmunol* **63**:193–205

30. Greer JM, Sobel RA, Sette A, Southwood S, Lees MB, Kuchroo VK (1996) Immunogenic and encephalitogenic epitope clusters of myelin proteolipid protein. *J Immunol* **156**:371–379

31. Targoni OS, Lehmann PV (1998) Endogenous myelin basic protein inactivates the high avidity T cell repertoire. *J Exp Med* **187**:2055–2063

32. Sundvall M, Jirholt J, Yang HT, Jansson L, Engstrom A, Pettersson U, Holmdahl R (1995) Identification of murine loci associated with susceptibility to chronic experimental autoimmune encephalomyelitis. *Nat Genet* **10**:313–317

33. Weissert R, Wallstrom E, Storch MK, Stefferl A, Lorentzen J, Lassmann H, Linington C, Olsson T (1998) MHC haplotype-dependent regulation of MOG-induced EAE in rats. *J Clin Invest* **102**:1265–1273

34. Goverman J Woods A, Larson L Weiner LP, Hood L, Zaller DM (1993) Transgenic mice that express a myelin basic protein-specific T cell receptor develop spontaneous autoimmunity. *Cell* **72**:551–560

35. Lafaille J, Nagashima K, Katsuki M, Tonegawa S (1994) High incidence of spontaneous autoimmune encephalomyelitis in immunodeficient anti-myelin basic protein T cell receptor transgenic mice. *Cell* **78**:399–408

36. Kumar V, Sercarz EE (1993) The involvement of T cell receptor peptide-specific regulatory CD4+ T cells in recovery from antigen induced autoimmune disease. *J Exp Med* **178**:909–916

37. Jiang H, Zhang SI, Pernis B (1992) Role of CD8+ T cells in murine experimental allergic encephalomyelitis. *Science* **256**:1213–1215

38. Kozovska MF, Yamamura T, Tabira T (1996) T-T cell interaction between CD4-CD8- regulatory T-cells and T cell clones presenting TCR peptide. Its implications for TCR vaccination against experimental autoimmune encephalomyelitis. *J Immunol* **157**:1781–1790

39. Huitinga I, Van Rooijen N, De Groot CJA, Uitdehaag BMJ, Dijkstra CD (1990) Suppression of experimental allergic encephalomyelitis in Lewis rats aftre elimination of macrophages. *J Exp Med* **172**:1025–1033

40. Wolf SD, Dittel BN, Hardardottir F, Janaway CA (1996) Experimental autoimmune encephalomyelitis induction in genetically B-cell deficient mice. *J Exp Med* **184**:2271–2278

41. Kobayashi Y, Kaway K, Ito K, Honda H, Sobue G, Yoshikai Y (1997) Aggravation of murine experimental allergic encephalomyelitis by administration of T-cell receptor γδ-specific antibodies. *J Neuroimmunol* **73**:169–174

42. Zhang BN, Yamamura T, Kondo T, Fujiwara M, Tabira T (1997) Regulation of experimental autoimmune encephalomyelitis by natural killer (NK) cells. *J Exp Med* **186**:1677–1687

43. Naparstek Y, Ben-Nun A, Holoshitz J, Reshef T, Frenkel A, Rosenberg M, Cohen IR (1983) T lymphocyte line producing or vaccinating against autoimmune encephalomyelitis (EAE). Functional activation induces peanut agglutinin receptors and accumulation in the brain and thymus of line cells. *Eur J Immunol* **13**:418–423

44. Ufret-Vincenty RL, Quigley L, Tresser N, Pak SH, Gado A, Hausmann S, Wucherpfennig KW, Brocke S (1998) In Vivo Survival of Viral Antigen-specific T Cells that Induce Experimental Autoimmune Encephalomyelitis. *J Exp Med 1998* **188**:1725–1738

45. Brocke S, Gaur A, Piercy C, Gautam A, Gijbels K, Fathman GC, Steinman L (1993) Induction of relapsing paralysis in experimental autoimmune encephalomyelitis by bacterial superantigen. *Nature* **365**:642–644

46. Padovan E, Giachino C, Cella M, Valitutti S, Acuto O, Lanzavecchia A (1995) Normal T lymphocytes can express two different T cell receptor β chains: implications for the mechanism of allelic exclusion. *J Exp Med* 181:1587–1591

47. Cannella B, Cross AH, Raine CS (1990) Upregulation and coexpression of adhesion molecules correlates with experimental autoimmune demyelination in the central nervous system. *J Exp Med* 172:1521–1524

48. Myers KJ, Dougherty JP, Ron Y (1993) In vivo antigen presentation by both brain parenchymal cells and hematopoietically derived cells during the induction of experimental autoimmune encephalomyelitis. *J Immunol* 15:2252–2260

49. Traugott U, McFarlin DE, Raine CS (1986) Immunopathology of the lesion in chronic relapsing experimental autoimmune encephalomyelitis in the mouse. *Cell Immunol* 99:394–410

50. Olsson T (1994) Role of cytokines in multiple sclerosis and experimental autoimmune encephalomyelitis. *Eur J Neurol* 1:7–19

51. Selmaj KW, Raine CS (1988) Tumor necrosis factor mediates myelin and oligodendrocyte damage in vitro. *Ann Neurol* 23:339–46

52. Compston DA, Scolding NJ (1991) Immune-mediated oligodendrocyte injury. *Ann N Y Acad Sci* 633:196–204

53. Conboy YM, DeKruyff R, Tate KM, Cao ZA, Moore TA, Umetsu DT, Jones PP (1997) Novel genetic regulation of T helper 1 (Th1)/Th2 cytokine production and encephalitogenicity in inbred mouse strains. *J Exp Med* 185:439–451

54. Baron JL, Madri JA, Ruddle NH, Hashim G, Janaway CA (1993) Surface expression of α4 integrin by CD4 T cells is required for their entry into brain parenchyma. *J Exp Med* 177:57–68

55. Beraud E, Balzano C, Zamora AJ, Varriale S, Bernard D, Ben-Nun A (1993) Pathogenic and non-pathogenic T lymphocytes specific for the encephalitogenic epitope of myelin basic protein: functional characteristics and vaccination properties. *J Neuroimmunol* 47:41–54

56. Sun D, Wekerle H (1986) Ia-restricted encephalitogenic T lymphocytes mediating EAE lyse autoantigen-presentig astrocytes. *Nature* 320:70–72

57. Jeong MC, Itzikson L, Uccelli A, Brocke S, Oksenberg JR (1998) Differential display analysis of murine encephalitogenic mRNA. *Int Immunol* 10:1819–1823

58. Mustafa M, Vingsbo C, Olsson T, Ljungdhal A, Hojeberg B, Holmdahl R (1993) The major histocompatibility complex influences myelin basic protein 68–86-induced T-cell cytokine profile and experimental autoimmune encephalomyelitis. *Eur J Immunol* 23:3089–3095

59. Forsthuber T, Yip HC, Lehmann PV (1996) Induction of TH1 and TH2 immunity in neonatal mice. *Science* 271:1728–1730

60. Krakowski ML, Owens T (1997) The central nervous system environment controls effector CD4+ T cell cytokine profile in experimental allergic encephalomyelitis. *Eur J Immunol* 27:2840–2847

61. Aloisi F, Ria F, Penna G, Adorini L (1998) Microglia are more efficient than astrocytes in antigen processing and in Th1 but not Th2 cell activation. *J Immunol* 160:4671–4680

62. Falcone M, Bloom BR (1997) A T helper cell 2 (Th2) immune response against non-self antigens modifies the cytokine profile of autoimmune T cells and protects against experimental allergic encephalomyelitis. *J Exp Med* 185:901–907

63. Kuchroo VK, Das MP, Brown JA, Ranger AM, Zamvil SS, Sobel RA, Weiner HL, Nabavi N, Glimcher LH (1995) B7-1 and B7-2 costimulatory molecules activate differentially the Th1/Th2 developmental pathways: application to autoimmune disease therapy. *Cell* 80:707–718

64. Nicholson LB, Greer JM, Sobel RA, Lees MB, Kuchroo VK (1995) An altered peptide ligand mediates immune deviation and prevents autoimmune encephalomyelitis. *Immunity* 3:397–405

65. Weiner HL (1997) Oral tolerance: immune mechanisms and treatment of autoimmune diseases. *Immunol Today* 18:335–343

66. Racke MK, Bonomo A, Scott DE, Cannella B, Levine A, Raine CS, Shevach EM, Rocken M (1994) Cytokine-induced immune deviation as a therapy for inflammatory autoimmune disease. *J Exp Med* 180:1961–1966

67. Ruddle NH, Bergman CM, McGrath KM, Lingenheld EG, Grunnet ML, Padula SJ, Clark RB (1990) An antibody to lymphotoxin and tumor necrosis factor prevents transfer of experimental allergic encephalomyelitis. *J Exp Med* 172:1193–1200

68. Leonard JP, Waldburger KE, Goldman SJ (1995) Prevention of experimental autoimmune encephalomyelitis by antibodies against interleukin 12. *J Exp Med* 181:381–386

69. Genain CP, Abel K, Belmar N, Villinger F, Rosenberg DP, Linington C, Raine CS, Hauser SL (1996) Late complications of immune deviation therapy in a nonhuman primate. *Science* 274:2054–2057

70. Taupin V, Renno T, Bourbonniere L, Peterson AC, Rodriguez M, Owens T (1997) Increased severity of experimental autoimmune encephalomyelitis, chronic macrophage/microglial reactivity, and demyelination in transgenic mice producing

tumor necrosis factor-alpha in the central nervous system. *Eur J Immunol* 27:905–913

71. Frei K, Eugster HP, Bopst M, Constantinescu CS, Lavi E, Fontana A (1997) Tumor necrosis factor alpha and lymphotoxin alpha are not required for induction of acute experimental autoimmune encephalomyelitis. *J Exp Med* 185:2177–2182

72. Liu J, Marino MW, Wong G, Grail D, Dunn A, Bettadapura J, Slavin AJ, Old L, Bernard CC (1998) TNF is a potent anti-inflammatory cytokine in autoimmune-mediated demyelination. *Nat Med* 4:78–83

73. Ferber IA, Brocke S, Taylor-Edwards C, Ridgway W, Dinisco C, Steinman L, Dalton D, Fathman CG (1996) Mice with a disrupted IFN-gamma gene are susceptible to the induction of experimental autoimmune encephalomy elitis (EAE). *J Immunol* 156:5 7

74. Bettelli E, Das MP, Howard ED, Weiner HL, Sobel RA, Kuchroo VK (1998) IL-10 is critical in the regulation of autoimmune encephalomyelitis as demonstrated by studies of IL-10- and IL-4-deficient and transgenic mice. *J Immunol* 161:3299–3306

75. Sahrbacher UC, Lechner F, Eugster HP, Frei K, Lassmann H, Fontana A (1998) Mice with an inactivation of the inducible nitric oxide synthase gene are susceptible to experimental autoimmune encephalomyelitis. *Eur J Immunol* 28:1332–1338

76. Malipiero U, Frei K, Spanaus KS, Agresti C, Lassmann H, Hahne M, Tschopp J, Eugster HP, Fontana A (1997) Myelin oligodendrocyte glycoprotein-induced autoimmune encephalomyelitis is chronic/relapsing in perforin knockout mice, but monophasic in Fas- and Fas ligand-deficient lpr and gld mice. *Eur J Immunol* 27:3151–3160

77. Bontrop RE, Otting N, Slierendregt BL, Lanchbury JS (1995) Evolution of major histocompatibility complex polymorphisms and T-cell receptor diversity in primates. *Immunol Rev* 143:43–62

78. Massacesi L, Genain CP, Lee-Parritz D Letvin NL, Canfield D, Hauser SL. (1995) Chronic relapsing experimental autoimmune encephalomyelitis in new world primates. *Ann Neurol* 37:519–530

79. Uccelli A, Oksenberg JR, Jeong M, Genain CP, Rombos T, Jaeger E, Lanchbury J, Hauser SL (1997) Characterization of the TCRB chain repertoire in the New World monkey Callithrix jacchus. *J Immunol* 158:1201–1207

80. Antunes SG, de Groot NG, Brok H, Doxiadis G, Menezes AA, Otting N, Bontrop RE (1998) The Common Marmoset: a new world primate species with limited MHC class II variability. *Proc Natl Acad Sci U S A* 95:11745–11750

81. Genain CP, Nguyen MH, Letvin NL, Pearl R, Davis RL, Adelman L, Lees MB, Linington C, Hauser SL (1995) Antibody facilitation of multiple sclerosis-like lesions in a non-human primate. *J Clin Invest* 96:2966–2974

82. Hart BA, Bauer J, Muller HJ, Melchers B, Nicolay K, Brok H, Bontrop RE, Lassmann H, Massacesi L (1998) Histopathological characterization of magnetic resonance imaging-detectable brain white matter lesions in a primate model of multiple sclerosis: a correlative study in the experimental autoimmune encephalomyelitis model in common marmosets (*Callithrix jacchus*). *Am J Pathol* 153:649–663

83. Laman JD, van Meurs M, Schellekens MM, de Boer M, Melchers B, Massacesi L, Lassmann H, Claassen E, 't Hart BA (1998) Expression of accessory molecules and cytokines in acute EAE in marmoset monkeys (*Callithrix jacchus*). *J Neuroimmunol* 86:30–45

84. Genain CP, Roberts T, Davis R, Nguyen M, Uccelli A, Faulds D, Hoffman K, Timmel G, Li, Y, Ferrante R, Joshi N, Hedgpeth J, Hauser SL (1995) Prevention of autoimmune demyelination in non-human primates by a cAmp-specific phosphodiesterase inhibitor. *Proc Natl Acad Sci USA* 92:3601–3605

85. Shaw CM, Alvord EC, Hruby S (1988) Chronic remitting–relapsing experimental allergic encephalomyelitis induced in monkeys with homologous myelin basic protein. *Ann Neurol* 24:738–748

86. Massacesi L, Joshi N, Lee-Parritz D, Rombos A, Letvin NL, Hauser SL (1992) Experimental allergic encephalomyelitis in cynomolgus monkeys. Quantitation of T cell responses in peripheral blood. *J Clin Invest* 90:399–404

87. Slierendregt BL, Hall M, 't Hart B, Otting N, Anholts J, Verduin W, Claas F, Jonker M, Lanchbury JS, Bontrop RE (1995) Identification of an Mhc-DPB1 allele involved in susceptibility to experimental autoimmune encephalomyelitis in rhesus macaques. *Int Immunol* 7:1671–1679

88. Van Lambalgen R, Jonker M (1987) Experimental allergic encephalomyelitis in the rhesus monkey. Treatment of EAE with anti-T lymphocyte subset monoclonal antibodies. *Clin Exp Immunol* 67:305–312

89. Meinl E, Hoch RM, Dornmair K de Waal Malefyt R, Bontrop RE, Jonker M, Lassmann H, Hohlfeld R, Wekerle H, 't Hart B (1997) Differential encephalitogenic potential of myelin basic protein-specific T cell isolated from normal rhesus macaques. *Am J Pathol* 150:445–453

90. Levine S, Sowinski R (1973) Experimental allergic encephalomyelitis in inbred and outbred mice. *J Immunol* 110:139–43

Chapter 12

Transplantation of Oligodendrocyte Progenitor Cells in Animal Models of Leukodystrophies

Yoichi Kondo and Ian D. Duncan

Summary

Leukodystrophies represent a wide variety of hereditary disorders of the white matter in the central nervous system, where the patients, mostly in infancy or childhood, suffer from progressive and often fatal neurological symptoms due to either a delay or lack of myelin development or loss of myelin. As only supportive therapies are available for the majority of the leukodystrophies, replacing genetically defective oligodendrocytes with intact oligodendrocytes by transplantation has a potential as a curative therapy. Animal models of leukodystrophies have been valuable in developing effective strategies of myelin repair in human diseases. This chapter discusses the animal models of leukodystrophies and describes methods for (a) derivation of mouse oligodendrocyte progenitor cells (OPCs) in vitro as a source of donor myelin-forming cells and (b) transplantation of OPCs into the brain and spinal cord of mouse models of leukodystrophies.

Key words: Leukodystrophy, Myelin repair, Oligodendrocyte progenitor, Oligosphere, Transplantation

1. Introduction

The leukodystrophies are composed of a rare, but devastating, group of diseases of the white matter of the central nervous system (CNS). They encompass both diseases where myelination of the CNS delayed and absent, to those where myelination begins normally but subsequently demyelination ensues, resulting in delays in normal development of the affected children or the loss of neurologic function in previously normal children. Pelizaeus–Merzbacher disease (PMD) is the archetypical myelin development disorder, that along with its milder form, spastic paraplegia type II (SPG II), results in a range of severity of myelin abnormality from severe dysmyelination to myelin defects.

Neil J. Scolding and David Gordon (eds.), *Methods in Molecular Biology, Neural Cell Transplantation, vol. 549*
DOI: 10.1007/978-1-60327-931-4_12, © Humana Press, a part of Springer Science + Business Media, LLC 2009

The molecular defect in these patients is a mutation or duplication of the myelin protein *plp 1* gene *(1–5)*. In severely affected patients, practically no myelin is present throughout the CNS. In contrast to this developmental disorder, the X-linked recessive disease, adrenoleukodystrophy (ALD), and the autosomal recessive disorders, Krabbe's disease (globoid cell leukodystrophy) and metachromatic leukodystrophy (MLD), are characterized by massive demyelination. In the latter two disorders, demyelination also affects the peripheral nervous system (PNS). ALD can present in several forms, but in the most common acute cerebral form, the onset of disease and underlying demyelination begins around 5–6 years of age with a fulminant course of 2–3 years. In contrast, both Krabbe's disease and MLD usually develop in the first year of life with death in 2–3 years. In Krabbe's disease, there is a wide range of mutations in the galactocerebrosidase gene *(6)* and in MLD, mutations occur in the arylsulfatase A gene.

A common feature of all of these diseases is that no treatments are available that result in myelination or myelin restoration. Therapies such as dietary supplementations (e.g., Lorenzo's Oil in ALD) and bone marrow *(7)* or cord blood transplantation *(8)* have been useful in ameliorating or delaying the disease onset and severity, but these do not promote remyelination of the nervous system.

In efforts to repair the myelin defects in these diseases, certain animal models have and are proving invaluable. These models, collectively known as the myelin mutants, are a group of spontaneous mutations occurring in a range of species where the defect occurs in the gene(s) mutated or implicated in the leukodystrophies. In particular, there is a large group of X-linked mutants with mutations in the *plp 1* gene. Usefully, there is also a group of animals with mutations in the galactocerebrosidase gene that are excellent models of Krabbe's disease. In addition to these spontaneous mutants, transgenic or knockout mice or rats, in which the leukodystrophy genes have been targeted, are also useful in exploring repair of the CNS. These include mice with knockouts of the ALD gene(s) *(9–11)* or arylsulphatase A (MLD) *(12)*. These latter models demonstrate the biochemical effects of these diseases but not the severe phenotypic abnormalities. Nonetheless, they have been informative in gene therapy experiments in which there has been some evidence of myelin repair *(13)*.

In the mutants noted above, myelin repair has been attempted experimentally by the transplantation of myelin-producing cells. Greatest attention has been paid to the models of PMD, particularly the myelin deficient (*md*) rat *(14–17)*, and the canine mutant, the *shaking* pup *(18)*. These studies have shown that cell transplantation can result in large areas of myelination at the transplantation site, and that this can be scaled up from rodent species to larger animal models such as the *shaking* pup *(19)*.

The lack of the proteolipid protein in the *md* rat makes the evaluation of success of the transplant straightforward. Likewise, lack of myelin basic protein (MBP) in the *shiverer* (*shi*) mouse has led it to be the most used model to test the myelinating capacity of many cell types and both allo and xenografts *(20–23)*. While not a model of known leukodystrophy, the lack of myelin in *shi*, and the fact that it lives up to 4 months, makes it very useful. In contrast, the other murine leukodystrophy models have been less utilized to study myelin repair by cell transplantation.

The cells that have been transplanted into the mutants range from those of the oligodendrocyte lineage to Schwann cells. While the Schwann cells are of great interest, those of the oligodendrocyte lineage are likely to offer the greatest hope for large scale repair. A consensus of the experimental results to-date using such cells suggests that cells from the earlier stages of the lineage, either stem cells or progenitors, will be the most useful source of myelinating oligodendrocytes. In this chapter we will discuss these cells and the methods to produce them in large numbers for myelin repair.

2. Materials

2.1. OPC Culture from Mouse Neonates

1. Mouse neonates (postnatal days 0–3).
2. Dissecting stereomicroscope (e.g., Nikon SMZ series).
3. Four pairs of fine forceps (e.g., #5, Fine Scientific Tools, sterile).
4. Cultureware (T-25 flasks, 60-mm plastic dishes, 70-μm cell strainer, 50-mL conical tubes).
5. OPC Isolation Buffer: Hanks' balanced salt solution (HBSS; without Ca^{2+}, Mg^{2+}) containing 2 mM EDTA, 10 mM HEPES, and 1× penicillin/streptomycin.
6. Trypsin/EDTA: 0.25% trypsin/1 mM EDTA (Invitrogen).
7. DNase I (Worthington): Reconstitute at 2 mg/mL in HBSS. Filter-sterilize and store aliquots at –20°C.
8. Dulbecco's Modified Eagle's Medium (DMEM, Invitrogen).
9. Fetal Bovine Serum (FBS, HyClone).
10. Mouse oligosphere medium: Neurobasal-A medium containing 1× B27 supplement (Invitrogen), 2 mM glutamine, 2 μg/mL heparin, and 1× penicillin/streptomycin.
11. Murine recombinant basic-FGF and PDGF-AA (PeproTech). Adjust to 10 μg/mL in HBSS containing 2% bovine serum albumin (BSA) and store the aliquots at –20°C. Use up the thawed vial in 2 weeks.

2.2. Dissociation of Oligospheres

1. Fire-polished Pasteur pipette: Plug the end of 9-in. glass pipettes (Fisher Scientific) with cotton and autoclave. Create a small and smooth bore (430–450 μm diameter) by flaming the pipette tip with Bunsen burner. Pretreat the inner wall of pipette with HBSS/2% BSA – this prevents loss of spheres on the pipette wall. Alternatively, the pipettes can be coated with Sigmacote (Sigma) before autoclaving.

2.3. Transplantation

1. Recipient mice (e.g., twitcher mice, stock# 000845 and shiverer mice, stock# 001428, Jackson Laboratory).

2. Needle/pipette puller (David Kopf Instrument).

3. Surgical stereomicroscope (e.g., Nikon SMZ series).

4. Isoflurane anesthesia system (vaporizer, O_2 cylinder, etc.).

5. Hamilton syringe (10 μL, #80075).

6. Programmable syringe pump (GENIE, Kent Scientific).

7. Borosilicate glass capillary (1B100F-4, World Precision Instruments). Heat-pull the capillary and connect with the Hamilton syringe via PTFE tubing. Fill inside of the injection system with sterile saline and draw and eject a few microliters of HBSS/2% BSA to pretreat the inner wall of the needle.

8. 31-gauge insulin syringe (Becton Dickinson, Ultra-Fine II). Bend the needle tip with a needle holder, so the needle has approximately a 120° angle.

9. Heating pad (heating lamp is not recommended).

10. Microdrill and a 0.5-mm diameter steel burr (Fine Scientific Tools).

3. Methods

3.1. OPC Culture from Mouse Neonates

1. Spray 70% ethanol on the pups and decapitate them with sharp scissors. Under a dissection microscope in a laminar flow hood, remove the brain with two pairs of fine forceps and transfer it to 5 mL of ice-cold Isolation Buffer in a 60-mm culture dish. Using new fine forceps, remove the meninges and choroid plexus. In a new dish with 4.5 mL of Isolation Buffer, mince the brain into <1 mm³ blocks and transfer the tissue with the buffer into a 50-mL conical tube.

2. Add 0.5 mL of trypsin/EDTA and 50 μL of DNase I to the tube and incubate at 37°C for 20 min. Inactivate trypsin by adding 5 mL of DMEM/10% FBS and dissociate the tissue by pipetting with a 10-mL pipette up and down 10–15 times without making bubbles. Filtrate the cell suspension through a 70-μm cell strainer. Wash off the strainer by adding 10 mL of Isolation Buffer. Centrifuge at $200 \times g$, 7 min. Discard supernatant.

3. Resuspend the cells in 4 mL of mouse oligosphere medium and transfer the cells to a T-25 flask. Add basic-FGF and PDGF-AA at 20 ng/mL and 10 ng/mL, respectively. Culture the cells in a humidified incubator at 37°C, 5% CO_2. Exchange half of medium every 2–3 days. A number of spheres will form in 7 days. Cultures for transplantation are best used between 1 and 2 months (*see* **Fig. 1**; **Notes 2** and **3**.)

3.2. Dissociation of Oligospheres

1. Collect oligospheres in a 15-mL conical tube. Spin down the spheres at 200 × *g*, 3 min. Discard supernatant and wash the pellet with Isolation Buffer at 200 × *g*, 3 min. Discard the supernatant and add 2 mL of trypsin/EDTA. Incubate the tube in the water bath for 5 min at 37°C.

2. Triturate the spheres by pipetting up and down several times with a fire-polished Pasteur pipette connected to a Pipet-Aid dispenser (Drummond). Sit the cell suspension until unbroken spheres settle on the bottom (approximately 1 min). Transfer the supernatant to a new 15-mL tube containing 1 mL of DMEM/10% FBS. Add 2 mL of fresh trypsin/EDTA to the unbroken spheres.

3. Repeat **step 2** once or twice until the majority of spheres are dissociated into single cells. Count the number of cells in the single cell suspension and check their viability by Trypan blue staining.

4. Centrifuge the cell suspension at 200 × *g* for 7 min at 4°C. Discard the supernatant and adjust the cell concentration to

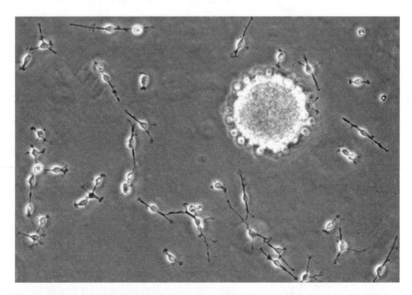

Fig. 1. Culture of oligodendrocyte progenitor cells (OPCs) as oligospheres. A floating oligosphere derived from the neonatal CNP-EGFP transgenic mouse *(24)* is illustrated, from which cells have migrated out and adhered to the plastic dish, with bi- or multipolar processes, the typical morphology of OPCs.

50,000–100,000 cells/μL with HBSS. Keep the cells in a 0.5-mL microcentrifuge tube on ice until and during transplantation surgery (*also see* **Notes 4–6**).

3.3. Transplantation into the Neonatal Brain

1. Wrap a mouse neonate in 4 in × 4 in. gauze and cover in crushed ice. The pup will be cryoanesthetized approximately in 2 min, showing pale skin, no movement, and no respiration. The time may need to be optimized for each laboratory.

2. Fill the heat-pulled needle with 2 μL of cell suspension by drawing a plunger. Hold the needle with your fingers and insert it into the brain at the level of bregma, 1 mm lateral from the midline, and 1.5 mm deep from the surface of skin. Marking the needle with a fine permanent pen at 1.5 mm from the tip will help control the depth of insertion. Inject the cells in 1 min using the programmable pump. Perform injection under the surgical microscope, so that the position of hand-held needle is properly fixed. Wait for 30 s before withdrawing the needle to prevent reflux of the injected cells. Uni- or bilateral injection can be performed. Warm up the pup on a heating pad. Return the pup to the mother after it starts breathing and moving.

3.4. Transplantation into the Brain of Young/Adult Mice

Intraparenchymal injection into the brain can be performed in young/adult mice (**Fig. 2**). Isoflurane gas (1.0–1.7% with 0.5 L/min O_2) is used for anesthesia. For adult mice, stereotactic frames are commercially available, which may yet be too large for small mice of 20–30 days old. For these mice, we use a hand-made Styrofoam mold to fix the head and hold the surface of skull perpendicular to the needle insertion. After a midline skin incision, the skull is exposed and a burr hole is made carefully using a microdrill with a 0.5-mm burr so as not to damage the dura matter. The coordinates to deliver the cells to the subcortical white matter will vary depending on the size of animals. For 21-day-old twitcher mice, we have been successful with the coordination of 0 mm anterior–posterior to the bregma, 2 mm lateral to the midline, and 2 mm ventral from the dura matter. After injection through the heat-pulled glass needle (hand-held or fixed on the stereotactic frame), the bur hole is filled with bone wax and skin incision closed.

3.5. Transplantation into the Dorsal White Matter of the Spinal Cord

1. Anesthetize a 3–4-week-old mouse with isoflurane gas and place it onto a frame. Because the target area is small and shallow from the cord surface, it is important to keep the cord from respiratory movement during injection by stabilizing the spinal column using a commercially available small-animal frame or one that is handmade. A pair of hooked forceps connected to a stable frame can maintain the stability of the spinal column of small mice.

2. Under the surgical microscope, make a dorsal midline skin incision and remove the spinous process at the thoracolumbar level. For a small pup, you can do so by cutting the lamina of vertebral arch with iris scissors (dedicated to this purpose);

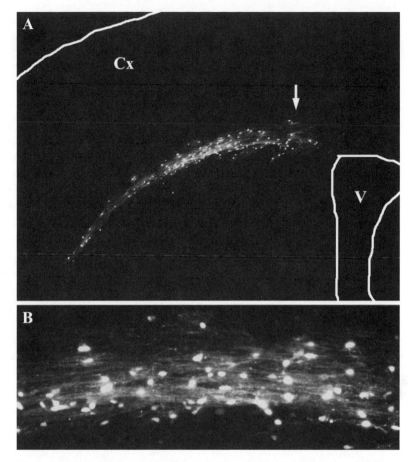

Fig. 2. Engrafted oligodendrocytes/myelin in the subcortical white matter of twitcher mouse brain. A twitcher mouse received intraparenchymal transplantation of OPCs derived from the CNP-EGFP transgenic mouse at P21 and was sacrificed 4 weeks after surgery. OPCs injected into the corpus callosum (*arrow*) spread along the white matter tract and differentiated into myelinating oligodendrocytes, expressing EGFP marker protein under the control of 2′,3′-cyclic nucleotide 3′-phosphodiesterase (CNP) promoter. Oligodendrocyte cell bodies show intense EGFP signals and the cell processes/myelin to a less extent. (b) is a higher magnification of the graft shown in (a). *Cx* Cerebral cortex and *V* Lateral ventricle.

while for a young/adult mouse, a microdrill with a 0.5-mm burr may be useful to cut the lamina.

3. Expose the dorsal column and clear the surface with a cotton applicator or a surgical spade. Do not damage the median dorsal spinal artery, or bleeding will require clearing of the surgical field. Alternatively, you may electrocauterize the artery, which will allow you to inject right on the midline. With a bent 31-gauge needle, cut a short length of the dura matter at the injection site.

4. Using a micromanipulator, insert the pulled glass needle in which 1 μL of cell suspension (50,000–100,000 cells) is

loaded, to the depth of 0.5 mm. Inject 1 μL at 0.2 μL/
min and withdraw the needle 1 min after the injection is
finished.

5. Close the skin incision and place the animal on the heating
   pad until recovery (*also see* **Note 7**).

3.6. Histological Evaluation

A critical issue will be to distinguish the donor-derived myelin/
oligodendrocyte from the endogenous cells and their myelin
sheaths. The simplest method is to use donor OPCs derived
from a transgenic mouse that expresses a marker protein such as
enhanced green fluorescent protein (EGFP) or β-galactosidase.
Choosing a promoter gene specific to oligodendrocyte lineage
cells helps identify the donor myelin/oligodendrocyte without
further histological processing (**Fig. 2**).

Choosing a myelin mutant as the recipient provides an advantage in identifying the donor-derived myelin. For example, the
shiverer mouse lacks compact myelin sheaths in the CNS due to a
deletion in the MBP gene. Therefore, immunolabeling for MBP
exclusively identifies the donor-derived myelin.

One-micrometer semi-thin sections for toluidine blue myelin
staining and ultrathin sections for electron microscopy can be
obtained from the tissue embedded in epon plastic. These methods
are valuable in detailed observation of donor-derived myelination
as well as pathological changes in axon/myelin (**Fig. 3**).

4. Notes

1. Although not mentioned throughout the text, animal care
   procedures (e.g., sterile techniques and pain control) must
   properly follow federal or national guidelines and of your
   institution.

2. OPCs can be derived from the rat using a medium containing 30% supernatant from the culture of B104 neuroblastoma
   cell line *(25)*. Oligospheres can form directly from the striatal
   tissue or can be induced from neurosphere (neural stem cell)
   cultures *(26)*.

3. Beyond 2 months of culture, growth of murine and canine
   (27) oligosphere cells slows down. The cells that migrate out
   from the sphere and adhere on the flask bottom often show
   abnormal morphologic changes such as fibroblast-like elongated processes, which may be attributed to the effect of FGF-2.
   These cells are not good source of myelin-forming cells for
   transplantation. In contrast, rat oligosphere cells may be used
   for a more extended time.

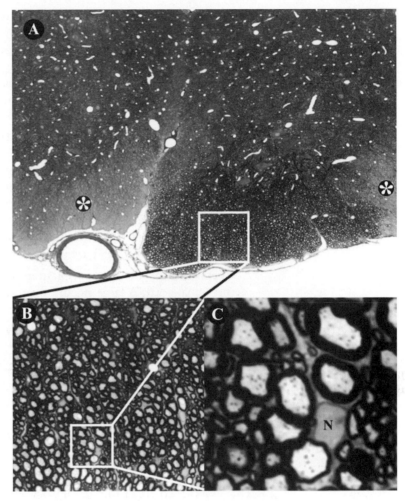

Fig. 3. Toluidine blue myelin staining on a 1-μm section of plastic embedded rat spinal cord. Shown is the spinal cord of the myelin deficient rat transplanted with OPCs derived from a normal rat brain. (a) The engrafted ventral white matter was extensively myelinated, yielding darker blue staining compared to the unmyelinated areas (*asterisk*). (b) Numerous myelinated axons are observed at a higher magnification. (c) Axons with various diameters are wrapped by different thickness of myelin. In general, donor-derived myelin is thinner than normally developed endogenous myelin. Here, however, myelin of large axons appears to be as thick as normal myelin. One nucleus (N) is possibly an oligodendrocyte.

4. Alternatively, oligospheres can be triturated in one step by using a fire-polished Pasteur pipette with a smaller bore size (<0.4 mm). This method works well for rat oligospheres. However, a large number of dead cells may be found in murine and canine spheres, because their OPCs appear to be more fragile to the shear stress.

5. Dissociated oligosphere cells can be plated to evaluate their in vitro differentiation. Cells are plated at 50,000 cells per 12-mm round glass coverslips coated with 0.1

mg/mL poly-L-ornithine and cultured with DMEM/FBS in a 24-well plate. When the medium contains 0.5–1% FBS, the cells differentiate into mature oligodendrocytes in 1 week, whereas the majority of cells become astrocytes with 10% FBS. In rats, more than 95% oligodendroglial differentiation can be achieved. However, the same method leads to as much as 20% oligodendroglial differentiation in murine and canine oligospheres, indicating that there are species differences in oligodendrocyte development.

6. In mice, it is our experience that oligosphere-derived OPCs do not show widespread migration into the brain parenchyma when transplanted into the third ventricle. Therefore, we prefer transplanting directly into the brain parenchyma *(23)*. As previously reported, rat OPCs in the ventricle can migrate extensively in the CNS parenchyma *(28)*.

7. The size of needle for transplantation is critical, especially in the spinal cord. A large needle wound induces inflammation, which may then reduce the survival of graft and/or damage host axons to be myelinated. However, a needle too small may be clogged with the dissociated OPCs clumped in the needle. We use a pulled glass needle with an outer diameter of 80–110 µm at the tip.

Acknowledgments

The work cited from our laboratory is currently supported by NMSS grant TR3761 and NIH NS055816. We acknowledge with gratitude the support of the Elisabeth Elser Doolittle Charitable Trust and the Oscar Rennebohm Foundation. We are grateful for past efforts of many previous colleagues.

References

1. Hodes ME, Pratt VM, Dlouhy SR. Genetics of Pelizaeus–Merzbacher disease. Dev Neurosci 1993;15(6):383–94.

2. Mimault C, Giraud G, Courtois V, et al. Proteolipoprotein gene analysis in 82 patients with sporadic Pelizaeus–Merzbacher Disease: duplications, the major cause of the disease, originate more frequently in male germ cells, but point mutations do not. The Clinical European Network on Brain Dysmyelinating Disease. Am J Hum Genet 1999;65(2):360–9.

3. Cailloux F, Gauthier-Barichard F, Mimault C, et al. Genotype–phenotype correlation in inherited brain myelination defects due to proteolipid protein gene mutations. Clinical European Network on Brain Dysmyelinating Disease. Eur J Hum Genet 2000;8(11):837–45.

4. Sistermans EA, de Coo RF, De Wijs IJ, Van Oost BA. Duplication of the proteolipid protein gene is the major cause of Pelizaeus–Merzbacher disease. *Neurology* 1998;50(6):1749–54.

5. Ellis D, Malcolm S. Proteolipid protein gene dosage effect in Pelizaeus–Merzbacher disease. *Nat Genet* 1994;6(4):333–4.

6. Wenger DA, Rafi MA, Luzi P. Molecular genetics of Krabbe disease (globoid cell leukodystrophy): diagnostic and clinical implications. *Hum Mutat* 1997;10(4):268–79.

7. Aubourg P, Blanche S, Jambaque I, et al. Reversal of early neurologic and neuroradiologic manifestations of X-linked adrenoleukodystrophy by bone marrow transplantation. *N Engl J Med* 1990;322(26):1860–6.

8. Escolar ML, Poe MD, Provenzale JM, et al. Transplantation of umbilical-cord blood in babies with infantile Krabbes disease. *N Engl J Med* 2005;352(20):2069–81.

9. Lu JF, Lawler AM, Watkins PA, et al. A mouse model for X linked adrenoleukodystrophy. *Proc Natl Acad Sci USA* 1997;94(17):9366–71.

10. Forss-Petter S, Werner H, Berger J, et al. Targeted inactivation of the X-linked adrenoleukodystrophy gene in mice. *J Neurosci Res* 1997;50(5):829–43.

11. Pujol A, Hindelang C, Callizot N, Bartsch U, Schachner M, Mandel JL. Late onset neurological phenotype of the X-ALD gene inactivation in mice: a mouse model for adrenomyeloneuropathy. *Hum Mol Genet* 2002;11(5):499–505.

12. Hess B, Saftig P, Hartmann D, et al. Phenotype of arylsulfatase A-deficient mice: relationship to human metachromatic leukodystrophy. *Proc Natl Acad Sci USA* 1996;93(25):14821–6.

13. Biffi A, Capotondo A, Fasano S, et al. Gene therapy of metachromatic leukodystrophy reverses neurological damage and deficits in mice. *J Clin Invest* 2006;116(11):3070–82.

14. Duncan ID, Hammang JP, Jackson KF, Wood PM, Bunge RP, Langford L. Transplantation of oligodendrocytes and Schwann cells into the spinal cord of the myelin-deficient rat. *J Neurocytol* 1988;17(3):351–60.

15. Tontsch U, Archer DR, Dubois-Dalcq M, Duncan ID. Transplantation of an oligodendrocyte cell line leading to extensive myelination. *Proc Natl Acad Sci USA* 1994;91(24):11616–20.

16. Brustle O, Jones KN, Learish RD, et al. Embryonic stem cell-derived glial precursors: a source of myelinating transplants. *Science* 1999;285(5428):754–6.

17. Rosenbluth J, Hasegawa M, Schiff R. Myelin formation in myelin-deficient rat spinal cord following transplantation of normal fetal spinal cord. *Neurosci Lett* 1989;97(1–2):35–40.

18. Archer DR, Cuddon PA, Lipsitz D, Duncan ID. Myelination of the canine central nervous system by glial cell transplantation: a model for repair of human myelin disease. *Nat Med* 1997;3(1):54–9.

19. Griffiths IR, Duncan ID, McCulloch M. Shaking pups: a disorder of central myelination in the spaniel dog. II. Ultrastructural observations on the white matter of the cervical spinal cord. *J Neurocytol* 1981;10(5):847–58.

20. Gumpel M, Baumann N, Raoul M, Jacque C. Survival and differentiation of oligodendrocytes from neural tissue transplanted into new-born mouse brain. *Neurosci Lett* 1983;37(3):307–11.

21. Baron-Van Evercooren A, Avellana-Adalid V, Ben Younes-Chennoufi A, Gansmuller A, Nait-Oumesmar B, Vignais L. Cell–cell interactions during the migration of myelin-forming cells transplanted in the demyelinated spinal cord. *Glia* 1996;16(2):147–64.

22. Windrem MS, Nunes MC, Rashbaum WK, et al. Fetal and adult human oligodendrocyte progenitor cell isolates myelinate the congenitally dysmyelinated brain. *Nat Med* 2004;10(1):93–7.

23. Kondo Y, Wenger DA, Gallo V, Duncan ID. Galactocerebrosidase-deficient oligodendrocytes maintain stable central myelin by exogenous replacement of the missing enzyme in mice. *Proc Natl Acad Sci USA* 2005;102(51):18670–5.

24. Yuan X, Chittajallu R, Belachew S, Anderson S, McBain CJ, Gallo V. Expression of the green fluorescent protein in the oligodendrocyte lineage: a transgenic mouse for developmental and physiological studies. *J Neurosci Res* 2002;70(4):529–45.

25. Avellana-Adalid V, Nait-Oumesmar B, Lachapelle F, Baron-Van Evercooren A. Expansion of rat oligodendrocyte progenitors into proliferative "oligospheres" that retain differentiation potential. *J Neurosci Res* 1996;45(5):558–70.

26. Zhang SC, Lundberg C, Lipsitz D, OConnor LT, Duncan ID. Generation of oligodendroglial progenitors from neural stem cells. *J Neurocytol* 1998;27(7):475–89.

27. Zhang SC, Lipsitz D, Duncan ID. Self-renewing canine oligodendroglial progenitor expanded as oligospheres. *J Neurosci Res* 1998;54(2):181–90.

28. Learish RD, Brustle O, Zhang SC, Duncan ID. Intraventricular transplantation of oligodendrocyte progenitors into a fetal myelin mutant results in widespread formation of myelin. *Ann Neurol* 1999;46(5):716–22.

Chapter 13

A Rat Middle Cerebral Artery Occlusion Model and Intravenous Cellular Delivery

Masanori Sasaki, Osamu Honmou, and Jeffery D. Kocsis

Summary

A useful experimental model to study the pathophysiology of cerebral ischemia without craniectomy is the middle cerebral artery occlusion (MCAO) model. In this model, an intraluminal suture is advanced from the internal carotid artery to occlude the base of the MCA. Standardized procedures in terms of suture size, animal weight, and the details of intraluminal suture insertion are well established. This procedure can produce reversible occlusion after insertion of the intraluminal suture for a specified period of time, or a permanent occlusion by leaving the suture in place. This model has been useful in the study of both the normal pathophysiology of cerebral ischemia and in assessing interventional therapeutic approaches for stroke therapy. One approach has been the intravenous delivery of bone marrow-derived mescenchymal stem cells at various times after MCAO. Histological and magnetic resonance imaging have been used to quantify infarction volume in this model system.

Key words: Middle cerebral artery occlusion, Stroke, TTC, Transplantation, Intravenous

1. Introduction

There are several model systems in rodent to induce reductions in global cerebral blood flow including multiple vessel occlusion and hypotension *(1, 2)*. Longa et al. *(3)* developed a much less invasive stroke model for either temporary or permanent regional ischemia, which is now commonly utilized in cerebral ischemia experiments in rodents. This model has allowed a number of advances in better understanding the pathophysiology of stroke *(4–6)* and in preclinical work showing efficacy of interventional approaches that reduce lesion volume and improve functional outcome. These approaches include pharmacological *(7, 8)* and cell-based *(9–13)* interventional strategies.

Neil J. Scolding and David Gordon (eds.), *Methods in Molecular Biology, Neural Cell Transplantation, vol. 549*
DOI: 10.1007/978-1-60327-931-4_13, © Humana Press, a part of Springer Science + Business Media, LLC 2009

Transplantation of mesenchymal stem cells (MSC) after cerebral ischemia in the middle cerebral artery occlusion (MCAO) model has been reported to reduce infarction size and ameliorate functional deficits in rodent cerebral ischemia models *(11, 13, 14)*. In addition, gene-modified human mesenchymal stem cells (hMSC) have greater therapeutic benefits on stroke following transplantation *(15, 16)*. The release of trophic factors from transplanted MSCs may contribute to the reduction in infarction size and to the recovery of function following ischemia in recipient animals *(10–14, 17–20)*. Several hypotheses to account for these therapeutic effects have been suggested, and current thinking is that neuroprotective and neovascularization, rather than neurogenesis are responsible for the early therapeutic effects. In this chapter, we describe a protocol for intraluminal suture induction of MCAO, an intravenous delivery method for cells and the use of **2,3,5**-triphenyltetrazolium chloride (TTC) staining and magnetic resonance imaging (MRI) for assessing lesion volume in the MCAO model.

2. Materials

2.1. Induction of the Middle Cerebral Artery Occlusion in the Rat

1. Animals: Male Sprague–Dawley rats (250–300 g)
2. Anesthesia: Ketamine (75 mg/kg) and xylazine (10 mg/kg)
3. Surgical tools: Surgical Blade (#10, Becton Dickinson, NJ, USA), Dumont forceps (#5, Fine Science Tools, CA, USA), Scissor, (Cohan-Vannas, Fine Science Tools, CA, USA), Scissor, (Fine Iris Scissor, Fine Science Tools, CA, USA), Retractor, (Agricola, Fine Science Tools, CA, USA), Needle holder, (Olsen-Hegar, Fine Science Tools, CA, USA), Bipolar coagulation unit (MACAN, IL, USA), Micro clamp (Micro Serrefine, Fine Science Tools, CA, USA).
4. Surgical supplies: Cotton Pellets (Richmond Dental, NC, USA), Gauze Sponges (Tyco Healthcare, MA, USA), Suture (4-0 Nylon monofilament, Sharpoint and Look, PA, USA), Surgical scrub: Betadine and 70% (v/v) ethanol, 0.9% (w/v) saline, sterile.
5. Filament for MCA occlusion: 3-0 surgical suture (Dermalon, Sherwood Davis and Geck, UK).
6. Surgical/Dissecting Microscope (10–50×).
7. Vital Signs parameters: Rectal temperature monitor (e.g., 43TD; Yellow Springs Instruments, Yellow Springs, OH), Thermal blanket (e.g., GAYMAR, NY, USA), Heart rate monitor (e.g., NPB-40; Nellcor, Pleasanton, CA).
8. Physiological monitoring system: PE-50 (Becton Dickinson, NJ, USA), Heart rate and blood pressure monitor system (e.g., Spike2, CED, UK).

2.2. Intravenous Delivery of the Transplanted Cells	1. Anesthesia, Surgical tools (same as in **Subheading 2.1**).
	2. PE-10 (Becton Dickinson, NJ, USA).
	3. 1-ml Syringe (Becton Dickinson, NJ, USA).

2.3. TTC Staining and Quantitative Analysis of Infarct Volume

1. 2% solution of 2,3,5-triphenyltetrazolium chloride (TTC) in normal saline, (Sigma, USA).
2. Sodium pentobarbital (50 mg/kg).
3. Microtome (Microslicer DTK-3000, Dosaka, Japan) or equivalent such as a Vibratome.
4. NIH image software (MD, USA).

2.4. MRI

1. 7-T, 18-cm-bore superconducting magnet (Oxford Magnet Technologies).
2. UNITYINOVA console (Oxford Instruments, Oxford, UK; and Varian, Inc., Palo Alto, CA).

3. Methods

The overall objective is to induce focal cerebral infarction in a defined hemisphere in brain in order to study biochemical changes following stroke and to assess therapeutic interventions. The basic approach is to insert a monofilament (nylon suture) through the external carotid artery (ECA) and then into the common carotid where the filament is fed into internal carotid artery (ICA) until it blocks the middle cerebral artery (MCA). The filament can then be left in place for a designated period of time followed by removal of the filament, or the filament can be cut and left in place for a permanent MCA occlusion. The rats can then be studied at various postsurgery recovery times after lesion induction. Moreover, interventional approaches such as intravenous infusion of MSCs can be carried out at defined postinfarction times. The duration of MCA occlusion, the time of experimental intervention (cell, drug delivery, etc.), and survival times can be altered to suit individual experimental designs.

Assessment of infarction volume is an important outcome measure as are more refined histological and behavioral analyses. Here we describe two basic approaches to assess gross lesion volume: TTC staining of brain and in vivo MRI. Other measures such as cerebral blood flow, behavioral analyses, and detailed immunohistochemical studies can be found in the papers cited in **Subheading 1**. A critical element in the MCAO model is consistency in lesion volume between animals. Selecting animals of similar weight and gender and standardizing surgical procedures are important to achieve lesion consistency. Additionally monitoring temperature, blood pH, PO_2, and PCO_2 throughout the surgery will allow selection of animals with a similar physiological response to the procedure.

3.1. Induction of the Middle Cerebral Artery Occlusion in the Rat

1. Male Sprague–Dawley rats weighing 250–300 g are anesthetized by intraperitoneal (i.p.) injection of a mixture of ketamine (75 mg/kg) and xylazine (10 mg/kg).

2. Rectal temperature is monitored and body temperature is maintained at 37°C with an infrared heat lamp and standard heating pad. The area over the throat and the inner thigh are shaved and scrubbed with betadine.

3. To monitor blood pH, PO_2, and PCO_2 with a heart rate and blood pressure monitor system throughout the surgery, the left femoral artery is cannulated with a PE-50 for measuring.

4. The right common carotid artery (CCA) is exposed by first making a midline incision through the skin over the ventral neck muscles. A retractor is positioned to reflect the digastric and sternomastoid muscles, and the omohyoid muscle is separated. The occipital artery branches of the ECA are then isolated and coagulated. See **Fig. 1** for schematic of arterial system.

5. Next, the superior thyroid and ascending pharyngeal arteries are blunt dissected and coagulated. The ECA is dissected further distally and coagulated along with the terminal lingual and maxillary artery branches. The ICA is isolated and carefully separated from the adjacent vagus nerve by blunt dissection. Further dissection will identify the ansa of the glossopharyngeal nerve at the origin of the pteryopalatine artery. This artery is a posteriorly directed extracranial branch of the ICA and is ligated with 7-0 nylon suture close to its origin. After this dissection and occlusion of arterial branches, the ICA is the only remaining intact extracranial branch of the CCA.

6. A 6-0 suture is then placed loosely around the ECA stump, and a curved microvascular clip is used to occlude circulation of the CCA and ICA by clamping at the junction where they meet the ECA stump.

7. A 5.0-cm length of 3-0 surgical suture (Dermalon, Sherwood Davis and Geck, UK) with the tip rounded by heating near a flame is advanced from the ECA into the lumen of the ICA until it blocks the origin of the MCA (20.0–22.0 mm of advancement) (*see* **Note 1**).

8. At a specified time after MCAO (usually 45–60 min), reperfusion is performed by withdrawal of the suture until the tip cleared the ICA (*see* **Note 2**). The suture can be left in place for permanent occlusion.

9. The incision is closed.

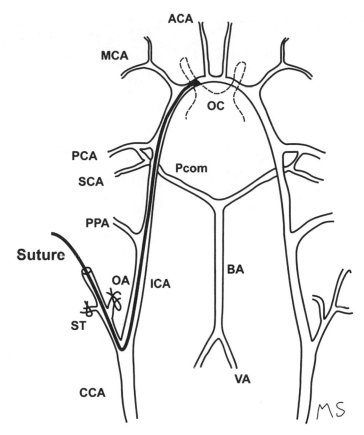

Fig. 1. Schematic showing suture insertion into the external carotid artery. *ACA* Anterior cerebral artery, *OC* Optic chiasm, *MCA* Middle cerebral artery, *PCA* Posterior cerebral artery, *SCA* Superior cerebellar artery, *Pcom* Posterior communicating artery, *PPA* Pterygopalatine artery, *BA* Basilar artery, *VA* Vertebral artery, *ST* Superior thyroid artery, *OA* Occipital artery, *ECA* External carotid artery, *ICA* Internal carotid artery, *CCA* Common carotid artery.

3.2. Intravenous Delivery of the Transplanted Cells Such as Mesenchymal Stem Cells

1. Cell suspension is prepared.

2. Rats are anesthetized with ketamine (75 mg/kg) and xylazine (10 mg/kg) i.p.

3. The left femoral vein is carefully exposed with dissecting microscope under sterile technique and cannulated using PE-10 tube connected to a 1-ml syringe. Injections of cells or control media can be directly delivered. This cannulation method assures that all of the cell suspension is injected.

4. The incision is closed.

3.3. TTC Staining and Quantitative Analysis of Infarct Volume (Fig. 2)

1. At any specified time after MCAO induction, the rats are deeply anesthetized with sodium pentobarbital (50 mg/kg, i.p.).

2. The brains are removed carefully and cut into coronal 1-mm sections using a tissue chopper or a Vibratome.

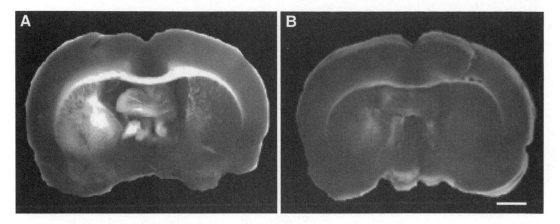

Fig. 2. Brain slices stained with 2,3,5-triphenyltetrazolium chloride (TTC) to visualize infarcted brain. TTC-stained brain slices from rats that were intravenously transplanted control media (**a**) and with autologous bone marrow cells at 6 h after MCAO (**b**). Note the larger infarction area particularly in basal ganglia nuclei in (**a**). Scale bar corresponds to 1.0 mm (modified from Iihoshi et al. *(11)*).

3. The fresh brain slices are immersed in a 2% solution of 2,3,5-triphenyltetrazolium chloride (TTC) in normal saline at 37°C for 30 min.

4. The cross-sectional area of infarction in each brain slice is examined and photographed with a dissection microscope and is measured using an image analysis software, NIH image.

5. The infarcted area appears white and normal brain red (*see* **Fig. 2**). The total infarct volume for each brain is calculated by summation of the infarcted area of all brain slices.

3.4. MRI (Fig.3)

1. Rats are anesthetized with ketamine (75 mg/kg) and xylazine (10 mg/kg) i.p.

2. Rat is placed in an animal holder/MRI probe apparatus and positioned inside the magnet. The animal's head is held in place inside the imaging coil.

3. MRI measurements are performed by using a 7-T, 18-cm-bore superconducting magnet (Oxford Magnet Technologies) interfaced to a UNITYINOVA console (Oxford Instruments, Oxford, UK; and Varian, Inc., Palo Alto, CA).

4. Diffusion weighted images (DWI) are obtained from a 1.0-mm-thick coronal section with a 3-cm field of view, TR = 3,000 ms, TE = 37 ms and are reconstructed with a 128×128 image matrix, b value = 966.

5. T_2-weighted images (T_2WI) are obtained from a 1.0-mm-thick coronal section with a 3-cm field of view, TR = 3,000 ms, TE = 30 ms and reconstructed with a 256×128 image matrix. Accurate positioning of the brain is performed to center the image slice 5 mm posterior to the rhinal fissure with the head of the rat held in a flat skull position.

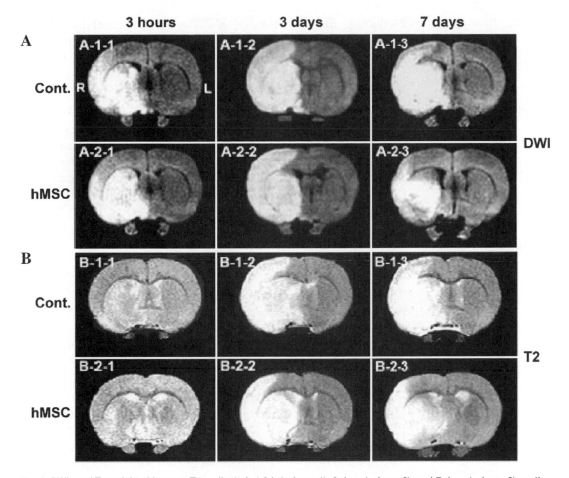

Fig. 3. DWI- and T$_2$-weighted images (T$_2$), collected at 3 h (column 1), 3 days (column 2), and 7 days (column 3) postlesion induction. The 3-h images were obtained just before intravenous cell or vehicle injection. Images from sham control and hMSC are shown in rows 1 and 2, respectively, for data sets in (**a**) DWI and (**b**) T$_2$. The infarction is observed as a high-intensity area of the ischemic side of the brain (modified from Liu et al. *(18)*).

6. The ischemic lesion area is calculated from both DWI and T$_2$WI with imaging software (Scion Image, Version Beta 4.0.2; Scion Corporation) (*see* **Note 3**).

4. Notes

1. Resistance is felt and a slight curving of the suture or stretching of the ICA is observed, indicating that the blunted tip of the suture has passed the MCA origin and reached the proximal segment of the anterior cerebral artery (ACA), which has a smaller diameter. At this point, the intraluminal suture has blocked the origin of the MCA, occluding all sources of blood flow from the ICA, ACA, and posterior cerebral artery (PCA).

2. During the induction of MCAO, physiological parameters (rectal temperature, blood pH, PO_2, PCO_2, blood pressure) should be maintained within normal ranges during surgery and transplantation procedures for all animals and should not statistically differ among the experimental groups.

3. For each MRI slice, the higher intensity lesions in DWI and T_2WI images where the signal intensity is 1.25 times higher than the counterpart in the contralateral brain lesion are marked as the ischemic lesion area, and infarct volume is calculated taking slice thickness (1 mm/slice) into account.

Acknowledgments

This work was supported in part by the Medical and Rehabilitation and Development Research Services of Department of Veterans Affairs, the National Institutes of Health (NS43432), and the National Multiple Sclerosis Society (RG2135; CA1009A10), the Paralyzed Veterans of America Research Foundation (2469), the Japanese Ministry of Education, Science, Sports and Culture (16390414), Mitsui Sumitomo Insurance Welfare Foundation, JST (Japan, Science and Technology Corporation) Innovation Plaza Hokkaido Project. The Center for Neuroscience and Regeneration Research is a collaboration of the Paralyzed Veterans of America and the United Spinal Association with Yale University.

References

1. Eklof B, Siesjo BK (1972) The effect of bilateral carotid artery ligation upon acid-base parameters and substrate levels in the rat brain. *Acta Physiol Scand*;**86**(4):528–38.

2. Salford LG, Plum F, Brierley JB (1973) Graded hypoxia-oligemia in rat brain. II. Neuropathological alterations and their implications. *Arch Neurol*;**29**(4):234–8.

3. Longa EZ, Weinstein PR, Carlson S, Cummins R (1989) Reversible middle cerebral artery occlusion without craniectomy in rats. *Stroke*;**20**: 84–91.

4. Garcia JH, Yoshida Y, Chen H, Li Y, Zhang ZG, Lian J, Chen S, Chopp M (1993) Progression from ischemic injury to infarct following middle cerebral artery occlusion in the rat. *Am J Pathol*;**142**:623–35.

5. Li Y, Sharov VG, Jiang N, Zaloga C, Sabbah HN, Chopp M (1995) Ultrastructural and light microscopic evidence of apoptosis after middle cerebral artery occlusion in the rat. *Am J Pathol*;**146**:1045–51.

6. Schaffer CB, Friedman B, Nishimura N, Schroeder LF, Tsai PS, Ebner FF, Lyden PD, Kleinfeld D (2006) Two-photon imaging of cortical surface microvessels reveals a robust redistribution in blood flow after vascular occlusion. *PLoS Biol*;**4**(2):e22.

7. Sun Y, Jin K, Xie L, Childs J, Mao XO, Logvinova A, Greenberg DA (2003) VEGF-induced neuroprotection, neurogenesis, and angiogenesis after focal cerebral ischemia. *J Clin Invest*;**111**(12):1843–51.

8. Ito D, Walker JR, Thompson CS, Moroz I, Lin W, Veselits ML, Hakim AM, Fienberg AA, Thinakaran G (2004) Characterization of stanniocalcin 2, a novel target of the mammalian unfolded protein response with cytoprotective properties. *Mol Cell Biol*;**24**(21):9456–69.

9. Chopp M, Li Y (2002) Treatment of neural injury with marrow stromal cells. *Lancet Neurol*;**1**(2):92–100.

10. Shen LH, Li Y, Chen J, Cui Y, Zhang C, Kapke A, Lu M, Savant-Bhonsale S, Chopp M (2007) One-year follow-up after bone marrow stromal cell treatment in middle-aged female rats with stroke. *Stroke*;**38**:2150–6.

11. Iihoshi S, Honmou O, Houkin K, Hashi K, Kocsis JD (2004) A therapeutic window for intravenous administration of autologous bone marrow after cerebral ischemia in adult rats. *Brain Res*;**1007**:1–9.

12. Nomura T, Honmou O, Harada K, Houkin K, Hamada H, Kocsis JD (2005) I.V. infusion of brain-derived neurotrophic factor gene-modified human mesenchymal stem cells protects against injury in a cerebral ischemia model in adult rat. *Neuroscience*;**136**:161–9.

13. Honma T, Honmou O, Iihoshi S, Harada K, Houkin K, Hamada H, Kocsis JD (2006) Intravenous infusion of immortalized human mesenchymal stem cells protects against injury in a cerebral ischemia model in adult rat. *Exp Neurol*;**199**(1):56–66.

14. Li Y, Chen J, Wang L, Lu M, Chopp M (2001) Treatment of stroke in rat with intracarotid administration of marrow stromal cells. *Neurology*;**56**:1666–72.

15. Kurozumi K, Nakamura K, Tamiya T, Kawano Y, Kobune M, Hirai S, et al. (2004) BDNF gene-modified mesenchymal stem cells promote functional recovery and reduce infarct size in the rat middle cerebral artery occlusion model. *Mol Ther*;**9**:189–97.

16. Hamada H, Kobune M, Nakamura K, Kawano Y, Kato K, Honmou O, Houkin K, Matsunaga T, Niitsu Y (2005) Mesenchymal stem cells (MSC) as therapeutic cytoreagents for gene therapy. *Cancer Sci*;**96**(3):149–56.

17. Mahmood A, Lu D, Wang L, Chopp M (2002) Intracerebral transplantation of marrow stromal cells cultured with neurotrophic factors promotes functional recovery in adult rats subjected to traumatic brain injury. *J Neurotrauma*;**19**(12):1609–17.

18. Liu H, Honmou O, Harada K, Nakamura K, Houkin K, Hamada H, Kocsis JD (2006) Neuroprotection by PlGF gene-modified human mesenchymal stem cells after cerebral ischaemia. *Brain*;**129**(Pt 10):2734–45.

19. Horita Y, Honmou O, Harada K, Houkin K, Hamada H, Kocsis JD (2006) Intravenous administration of glial cell line-derived neurotrophic factor gene-modified human mesenchymal stem cells protects against injury in a cerebral ischemia model in the adult rat. *J Neurosci Res*;**84**(7):1495–504.

20. Ukai R, Honmou O, Harada K, Houkin K, Hamada H, Kocsis JD (2007) Mesenchymal stem cells derived from peripheral blood protects against ischemia. *J Neurotrauma*;**24**(3):508–20.

Chapter 14

MR Tracking of Stem Cells in Living Recipients

Eva Syková, Pavla Jendelová, and Vít Herynek

Summary

Noninvasive cellular imaging allows the real-time tracking of grafted cells as well as the monitoring of their migration. Several techniques for in vivo cellular imaging are available that permit the characterization of transplanted cells in a living organism, including magnetic resonance imaging (MRI), bioluminescence, positron emission tomography, and multiple photon microscopy. All of these methods, based on different principles, provide distinctive, usually complementary information. In this review, we will focus on cell tracking using MRI, since MRI is noninvasive, clinically transferable, and displays good resolution, ranging from 50μm in animal experiments up to 300μm using whole body clinical scanners. In addition to information about grafted cells, MRI provides information about the surrounding tissue (i.e., lesion size, edema, inflammation), which may negatively affect graft survival or the functional recovery of the tissue. Transplanted cells are labeled with MR contrast agents in vitro prior to transplantation in order to visualize them in the host tissue. The chapter will focus on the use of superparamagnetic iron oxide nanoparticles (SPIO), because they have strong effects on T2 relaxation yet do not affect cell viability, and will provide an overview of different modifications of SPIO and their use in MR tracking in living organisms.

Key words: Contrast agets, Nanoparticles, Photochemical lesion, Spinal cord injury, Cell labeling, Stroke, Cellular imaging

1. Introduction

Stem and progenitor cells are being explored in regenerative medicine for the cell therapy of disorders of the central nervous system *(1–9)*. Crucial to the future success of cell transplantation in the clinical setting is the ability of transplanted cells to migrate from the site of transplantation to the lesioned area and to survive, differentiate, and replace lost nerve cells or produce growth factors and cytokines for a prolonged period of time in order to

Neil J. Scolding and David Gordon (eds.), *Methods in Molecular Biology, Neural Cell Transplantation vol. 549*
DOI: 10.1007/978-1-60327-931-4_14, © Humana Press, a part of Springer Science+Business Media, LLC 2009

enhance the patient's regenerative potential. Visualizing transplanted cells in vivo is essential for preclinical studies in rodents and potentially also in humans. The magnetic tracking of cells appears to be a valuable tool for such studies since MR imaging may serve to study cell migration to lesions, its time course, and how long the cells persist in the target region. Such information could help to elucidate the time window during which the transplantation of therapeutic cells may be clinically effective, the number of cells needed, and the optimal method of their administration. In addition to monitoring their migration, noninvasive cellular imaging allows the real-time tracking of grafted cells as well. Several techniques for in vivo cellular imaging are now available that allow the characterization of transplanted cells in a living organism, including magnetic resonance imaging (MRI), bioluminescence, positron emission tomography, and multiple photon microscopy. All of these imaging methods, based on different principles, provide distinctive, usually complementary information. In this chapter, we will focus on cell labeling for MRI, since MRI is noninvasive, clinically transferable and displays good resolution, ranging from 50μm in animal experiments up to 300μm when using whole body clinical scanners. In addition to information about grafted cells, MRI can provide information about the surrounding tissue (i.e., lesion size, edema, or inflammation), which may have a negative effect on graft survival or the functional recovery of the tissue.

1.1. Cellular Imaging

Magnetic resonance imaging is used in clinical practice because it provides good soft tissue contrast, based on the differential distribution of hydrogen atoms (^{1}H) in particular tissues. As hydrogen ions are present in water molecules, MRI images visualize the distribution of water in the extracellular and intracellular compartments of different types of tissues. MRI contrast agents alter the MR signal by changing the relaxivity of ^{1}H atoms in their vicinity. Changes in relaxivity are considered to affect either the spin–lattice relaxation (so-called T1 relaxation) or spin–spin relaxation (known as T2 relaxation). In an actual experimental system, T2 is reduced to T2* due to inhomogeneities in the magnetic field gradient. The relaxation rates R1 and R2 are the inverse of the relaxation times and express the effectiveness of the contrast agent. MR contrast agents contain metal ions that define their relaxation properties. Paramagnetic metals such as gadolinium, iron, and manganese mainly affect T1 relaxation, whereas superparamagnetic iron oxide (SPIO) nanoparticles predominantly reduce T2 and T2*. For cellular MRI, cells need to be labeled with an MR contrast agent in order to visualize them in the host tissue. Contrast agents suitable for cell labeling must not only have high relaxation rates, but they also must not affect the cells' viability or their ability to differentiate and migrate toward the target tissue. In addition, the contrast agent must either incorporate into the cell cytoplasm (intracellular cell labeling) or bind to the cell surface (extracellular cell labeling).

1.1.1. Intracellular Cell Labeling

Superparamagnetic iron oxide nanoparticles (SPIO) were introduced as contrast agents for cell labeling shortly after the use of gadolinium-chelates *(10)*. They are currently the preferred choice mainly because of the following properties (a) they provide the most change in signal per unit (size) of metal, in particular on T2*-weighted images; (b) they are composed of iron, which is biocompatible and can thus be recycled by cells using normal biochemical pathways for iron metabolism; (c) their surface coating allows the chemical linkage of functional groups and ligands; and (d) they can be easily detected by light and electron microscopy.

SPIO require stabilization in order to prevent aggregation. Most commonly this is accomplished by a surface coating of dextran. Dextran-coated iron oxide nanoparticles (DCSPIO) include the products Feridex® and Endorem® (**Fig. 1a**), the ultra-small SPIO Combidex® and Sinerem®, monocrystalline SPIO, and cross-linked SPIO. Endorem is available in the form of an aqueous colloid. Iron crystal size varies over a range of 4.3–5.6nm; the whole particle size is 120–150nm *(12)*. The particles are characterized by high relaxivities of the same order of magnitude as those of monocrystal iron-oxide nanoparticle-based contrast agents; however, the relaxivity related to iron content is lower, because iron does not create monocrystals and thus the superparamagnetic moment is lower. Nevertheless, the contrast agent Endorem can be easily incorporated by endocytosis, and all the cells survive and further divide in vitro. In experiments conducted in our laboratory, cell cultures of human or rat mesenchymal stem cells (hMSCs, rMSCs), rat olfactory ensheathing cells (rOECs), or human chondrocytes were incubated 2–3 days in media containing Endorem. Additionally, feeder-free D3 mouse embryonic stem cells (ESCs) transfected with the pEGFP-C1 vector were labeled with Endorem (112.4mg/ml) for three passages *(13)*. Nanoparticles were detected by staining for iron (Prussian blue; **Fig. 1b**). Transmission electron microscopy confirmed the presence of iron-oxide particles inside the cells, observed as membrane-bound clusters within the cell cytoplasm (**Fig. 1c**). On the day that nanoparticles were withdrawn, the efficiency of MSC labeling (i.e., how many cells out of the total number of analyzed cells were labeled) was 50–70%. The average amount of iron present in rat MSCs determined by spectrophotometry after mineralization of iron-labeled cell suspensions was 17pg/cell.

Another labeling approach is based on combining a commercially available dextran-coated SPIO, such as Feridex® or Sinerem®, and a commercially available transfection agent, for

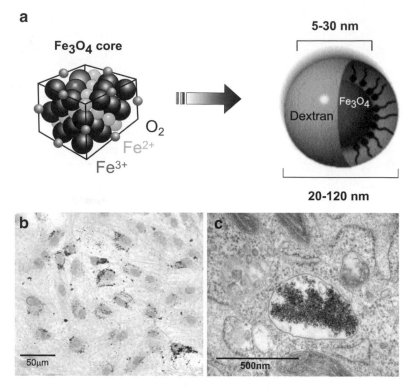

Fig. 1. Construction of iron-oxide nanoparticles and MSC labeling with iron-oxide nanoparticles. (**a**) The contrast agent Endorem consists of a superparamagnetic Fe_3O_4 core that is coated by a dextran shell. (**b**) Cells in culture labeled with BrdU (dark nuclei), containing superparamagnetic nanoparticles (*black dots*). (**c**) Transmission electron microphotograph of a cluster of iron nanoparticles surrounded by a cell membrane (modified from *(11)*).

example Superfect™, poly-L-lysine, Lipofectamin, or Fugene™. Transfection agents effectively transport nanoparticles into cells through electrostatic interactions. However, each combination of transfection agent and dextran-coated SPIO nanoparticle has to be carefully titrated and optimized for different cell cultures, since lower concentrations of the transfection agent may result in insufficient cellular uptake, whereas higher concentrations may induce the precipitation of complexes or may be toxic to the cells *(14)*. To overcome these drawbacks, new polycation-bound superparamagnetic iron oxide (PC-SPIO) nanoparticles have been developed *(15)*. PC-SPIO nanoparticles (**Fig. 2a**) combine the advantages of a low concentration of iron in the cell culture media (15.4 µg/ml) with facilitated uptake without the further use of any additives (transfection agents or lipofection) *(17)*. A comparison of standard dextran-coated SPIO and PC-SPIO showed that labeling cells with PC-SPIO nanoparticles was more efficient than labeling with Endorem, i.e., more cells out of the total number of analyzed cells were labeled with PC-SPIO nanoparticles than with

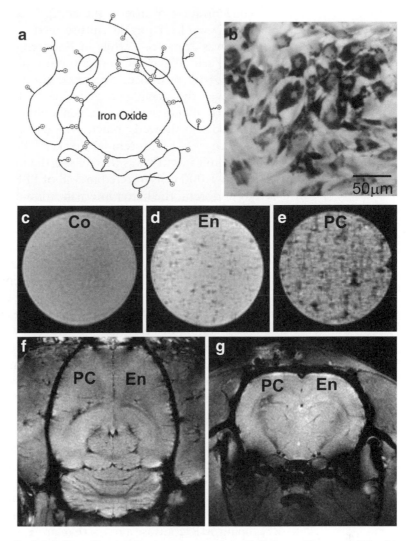

Fig. 2. Labeling with polycation-bound superparamagnetic iron oxide nanoparticles (**a**) Schematic illustration of a polycation-bound superparamagnetic iron oxide nanoparticle. (**b**) PC-SPIO-labeled MSCs in culture stained with Prussian blue. (**c–e**) MR images of phantoms formed by a set of test tubes containing a suspension of human MSCs in 0.5-ml gelatin. The cells were unlabeled (control, Co) (**C**) or labeled with Endorem (**d**) or with PC-SPIO nanoparticles (**e**). Phantoms contained 10,000 cells, corresponding to 0.4 cells per image voxel. (**f**) Axial and coronal (**g**) MR images of a rat brain with 1,000 cells labeled with PC-SPIO nanoparticles implanted in the left hemisphere (PC) and 1,000 Endorem-labeled cells implanted in the right hemisphere (En). MR images were taken 3 days after implantation (modified from *(16)*).

Endorem (**Fig. 2b**). The PC-SPIO nanoparticle suspension was used at a much lower iron concentration per milliliter of culture media (15.4 μg/ml) than with Endorem, for which 112.4 μg/ml in the culture media was used, and the cells were incubated with both labels for 3 days. In PC-SPIO-labeled cells, even though the concentration of iron in the culture media was ten times lower (15.4 vs. 112.4 μg/ml culture media), the average amount of iron was 38pg/cell, while in Endorem-labeled cells it was only 17pg/cell. When the concentration of Endorem was decreased to 15.4 μg of iron in 1ml of culture media, viability increased; however, the labeling efficiency fell to 40%.

MR images of phantoms containing suspensions of PC-SPIO-labeled cells showed a much stronger hypointense signal than did MR images of Endorem-labeled cells (**Fig. 2c–e**). The MR detection limit in vitro was 0.4 cells in the image voxel, while in vivo it was 1,000 cells injected in 5 μl of PBS (**Fig. 2f, g**). Compared to Endorem, the better internalization of PC-SPIO particles into the cells enables their easier MRI detection and tracking in the tissue after transplantation.

Another example of a suitable surface modification to enhance nanoparticle transport into the cells is coating with D-mannose. The cell surface is known to possess receptors for mannose *(18)*, and thus the coated nanoparticles can be easily internalized. New surface-modified iron oxide nanoparticles were developed by the precipitation of Fe(II) and Fe(III) salts with ammonium hydroxide followed by the oxidation of the precipitated magnetite with sodium hypochlorite with the subsequent addition of D-mannose solution *(19)*. The efficiency of MSC labeling was about 80%, and cells labeled with d-mannose-coated nanoparticles possessed very high relaxivity ($12.1s^{-1}$/million of cells/ml) when compared to cells labeled with Endorem ($1.24s^{-1}$/million of cells/ml). The average amount of iron as determined by spectrophotometry after mineralization was 51.7pg of iron per cell. Therefore, D-mannose-modified iron oxide nanoparticles are another promising tool for labeling living cells for diagnostic and therapeutic applications in cell-based therapies. Other strategies for developing MR contrast agents that are easy to detect include the utilization of viral protein cages *(20)* or the use of internalizing monoclonal antibodies *(21)*. Cells in suspension can be labeled using magnetoelectroporation *(22)*.

1.1.2. Extracellular Cell Labeling

The disadvantage of an intracellular label is that it can affect cell metabolism and subsequently cell viability. In addition, these labels are rather nonspecific; they can be loaded by virtually any cell present in the medium. Particles that do not internalize do not affect cell viability, and keeping cells in culture is not required for cell labeling; however, their attachment to the outer cell membrane is likely to interfere with cell surface interactions, and they may easily detach from the membrane or be transferred to other cells. Long-term follow up is therefore not possible.

The possibility of labeling only selected cell types would be very useful. Cells labeled and separated by means of immunomagnetic selection would not require in vitro culturing, since the label is attached during separation. This would allow the immediate clinical use of labeled cells. The first experiments using contrast agents bound to an antibody that can specifically bind to a single cell type were performed by Bulte and coworkers *(23)*. They described experiments with human lymphocytes labeled with biotinylated antilymphocyte-directed monoclonal antibodies, to which streptavidin and subsequently biotinylated dextran-magnetite particles were coupled. We tested another type of specific cell labeling using commercially available cell isolation kits for the magnetic separation of CD34$^+$ cells *(24)*. CD34$^+$ cells are known as hematopoietic progenitor cells. The cells were separated by means of immunomagnetic selection with anti-CD34 antibodies. For sorting, a superparamagnetic iron oxide core coated with a polysaccharide is linked to an antibody that binds to the respective cell type (**Fig. 3a**). The size of the label is comparable to commonly used superparamagnetic MR contrast agents, thus it can provide sufficient contrast on MR images.

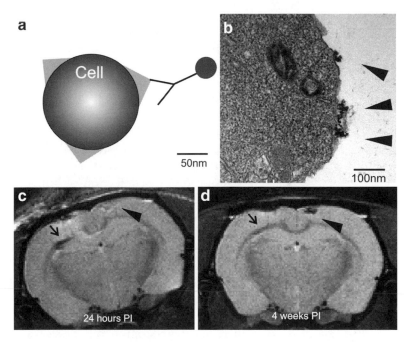

Fig. 3. Scheme of Microbeads and labeling with an extracellular magnetic label. (**a**) Scheme of MicroBeads: A superparamagnetic iron oxide core coated with a polysaccharide is linked to an antibody, which in turn is bound to a cell via an antigen–antibody complex. (**b**) Transmission electron microphotograph of MicroBeads binding to the cell surface. (**c**) A cell implant (in the hemisphere contralateral to the lesion; *arrow*) seen as a hypointense area on MR images 24 h postimplantation (PI). The lesion (*arrowhead*) remains hyperintensive. (**d**) A decreased signal (*arrow*) in the contralateral hemisphere, while a hypointense signal in the lesion (*arrowhead*) is visible 4 weeks after grafting (modified from *(24)*).

Human CD34+ cells from G-CSF mobilized peripheral blood were selected by CliniMACS CD34+ Selection Technology (Miltenyi). On electron microscopy images, we determined that the nanoparticles remained bound to the cell surface, and we observed several iron labels attached to the cell surface (**Fig. 3b**). The average iron content per cell, determined by spectrometry, was 0.275 pg. This value is lower by two orders of magnitude than in the case of intracellular cell labels; nevertheless, it still provides reasonable MR contrast (**Fig. 3c, d**; *(24)*).

1.2. MR Tracking of Implanted Cells in the Injured Brain

The use of stem cell therapy in stroke has been extensively examined. Endorem-labeled hMSCs, Endorem/BrdU co-labeled rMSCs, or Endorem-labeled eGFP ESCs were grafted into rats (Wistar, males, 2–3 months old) with a cortical photochemical lesion – an experimental model of stroke *(11, 13, 25)*. The cells were grafted either intracerebrally into the hemisphere contralateral to the lesion or intravenously into the femoral vein. Rats with grafted stem cells were examined weekly for a period of 3–7 weeks posttransplantation using a 4.7 T Bruker spectrometer. Single sagittal, coronal, and transversal images were obtained by a fast gradient echo sequence for localizing subsequent T2-weighted transversal images measured by a standard turbospin echo sequence. The lesion was visible on MR images already 2h after induction as a hyperintense signal and remained visible during the entire measurement period. One week after grafting, a hypointense signal was found in the lesion, which intensified during the second and third weeks regardless of the route of administration (**Fig. 4a, b**); its intensity corresponded to Prussian blue staining, anti-BrdU staining (**Fig. 4c–e**), or GFP labeling *(13, 25)*. Human MSCs were also demonstrated in the lesion by PCR detection of human DNA (a human-specific 850bp fragment of α-satellite DNA from human chromosome 17). Similar results were obtained after the injection of ESCs into rats with transient focal cerebral ischemia; the cells migrated toward the lesion *(26)*. Migration of gadolinium/rhodamine-labeled neural progenitors invading the penumbra was observed in a middle cerebral artery occlusion stroke model *(27)*).

1.3. MR Tracking of Implanted Cells in the Injured Spinal Cord

Spinal cord injury (SCI) invariably results in the loss of neurons and axonal degeneration at the lesion site, leading to severe functional impairment, paraplegia, or tetraplegia. Currently, there is no effective treatment for SCI. Standard treatment consists of stabilization and a conservative therapy using high doses of methylprednisolone; long-term therapy after SCI focuses on rehabilitation, pain relief, spasticity treatment, and the prevention of complications. The clinically most relevant model of SCI is a balloon-induced compression lesion *(28)*. A 2-French Fogarty

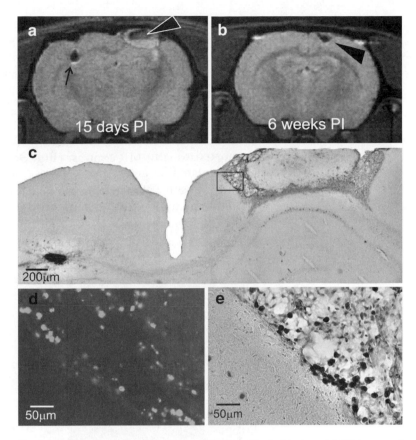

Fig. 4. Implantation of MSCs into brains with photochemical lesions. (**a**) MR images of a cortical photochemical lesion and MSCs implanted into the contralateral hemisphere. The cell implant (*black arrow*) in the hemisphere contralateral to the lesion and the lesion itself are seen as hypointensive areas (*black arrowhead*) 15 days after grafting. (**b**) MR images of a cortical photochemical lesion and MSCs injected into the femoral vein. A hypointensive signal (*black arrowhead*) observed 6 weeks after intravenous injection. (**c**) Prussian blue staining of an injection site in the contralateral hemisphere and a photochemical lesion, 4 weeks after grafting. (**d**) Higher magnification microphotograph (inset) anti-BrdU staining showing BrdU-positive MSCs in the lesion and (**e**) Prussian blue-stained MSCs along the left edge of the photochemical lesion (modified from *(25)*).

catheter is inserted into the dorsal epidural space through a small hole made in the Th10 vertebral arch. A spinal cord lesion is made by balloon inflation (volume 15 μl) at the Th8–Th9 spinal level. Inflation for 5min produces paraplegia and is followed by gradual recovery.

MSCs labeled with Endorem were injected intravenously into the femoral vein 1 week after a transversal spinal cord lesion *(11)*. MR images were taken ex vivo 4 weeks after cell implantation using a standard whole body resonator. MR images of longitudinal spinal cord sections from lesioned nongrafted animals showed the lesion cavity as inhomogeneous tissue with a strong hyperintensive signal (**Fig. 5a**). Lesions of grafted animals were seen as dark hypointense areas (**Fig. 5c**), while lesions of animals injected with nanoparticles in PBS were visible as weak hypointense signals (**Fig. 5e**). Histological evaluation confirmed only a few iron-containing cells in lesioned, control animals (**Fig. 5b**), strong positivity for iron in grafted animals (**Fig. 5d**), and a few weakly stained Prussian blue-positive cells in animals injected intravenously with PBS containing nanoparticles (**Fig. 5f**). Compared to control rats, in grafted animals the lesion, which was populated by grafted MSCs, was considerably smaller, suggesting a positive effect of the MSCs on lesion repair.

The migration of magnetodendrimer-labeled olfactory ensheathing glia (OEG) was also followed in the injured rat

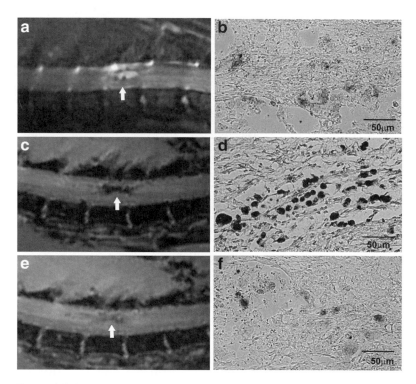

Fig. 5. MR images and Prussian blue staining of spinal cord compression lesions injected with saline, contrast agent, or nanoparticle-labeled MSCs. (**a**) Longitudinal MR image of a spinal cord lesion 5 weeks after induction. The formation of a lesion cavity is visible as a strong hyperintensive signal (*arrow*). (**b**) A few cells weakly stained for Prussian blue (*dark dots*) can be seen in a spinal cord lesion in an animal without implanted cells. (**c**) Longitudinal MR image 4 weeks after MSC grafting. The lesion with nanoparticle-labeled cells is visible as a dark hypointensive area (*arrow*). (**d**) Dense Prussian blue staining of a lesion populated with nanoparticle-labeled MSCs (*black dots*). (**e**) Longitudinal MR image 4 weeks after the injection of contrast agent. A weak hypointensive signal can be seen in the center of the lesion (*arrow*). (**f**) Weak Prussian blue positivity (*dark dots*) observed in a spinal cord lesion in an animal injected with contrast agent alone (adapted from *(29)*).

spinal cord. Labeled OEG were detectable in vivo by MR imaging for at least 2 months. Labeled OEG migrated in the normal spinal cord as demonstrated by MRI and histological markers. In contrast, OEG showed limited migration in transected spinal cords and were not able to cross the transection gap *(30)*.

So far, all the described studies employed USPIO or SPIO nanoparticles. Experiments with magnetic microspheres were performed in order to label Schwann cell transplants in a demyelinated spinal cord lesion *(31)*. The micron-sized cell label is much more readily detectable by MR and can be distinguished from the surrounding hemorrhage. A potential problem can be the limited uptake of these very large particles. However, the authors demonstrated that transplanted Schwann cells containing up to eight cytoplasmic microspheres were able to myelinate axons in vivo.

1.4. MR Tracking of Implanted Cells in Neurodegenerative Disorders

Multiple sclerosis is an immune-mediated demyelinating disease of the CNS causing significant disability in young adults. The beneficial effects of cell transplantation have now been shown in experimental autoimmune encephalomyelitis (EAE), an experimental model relevant to human multiple sclerosis *(32)*. The study demonstrated the ability to follow the migration of transplanted PLL-Feridex-labeled neural precursor cells by serial in vivo MR imaging in a mouse model of EAE. The transplanted cells responded to brain inflammation by migrating into the affected white matter tracts. The extent of migration correlated well with the clinical severity of the disease. Magnetic labels can be used for visualizing the multiple inflammatory foci that are found around blood vessels with a disrupted blood–brain barrier. These foci can be visualized by MRI following the intravenous injection of USPIO *(33–35)*, since the magnetic label can be either phagocytized by circulating monocytes or leak through the disrupted blood–brain barrier followed by uptake in resident brain microglia. MRI can also be used to study the contribution of T cells to lesion development *(36)*. The detection of macrophage trafficking in vivo can be generally used to predict the efficacy of anti-inflammatory drugs *(37)* or other treatments, potentially providing a marker of treatment efficacy, and thus may be used in future clinical applications.

1.5. MR Tracking of Implanted Cells in Brain Tumors

Recent evidence suggests that stem cells are useful delivery vehicles for brain tumor therapy *(38, 39)*. MR tracking was successfully used to demonstrate the incorporation of magnetically labeled bone marrow-derived precursor cells into the tumor vasculature as part of ongoing angiogenesis and neovascularization *(40)*. Glioma-bearing severe combined immunodeficient mice that received SPIO-labeled cells showed hypointense regions within the tumor area. Histology showed iron-labeled cells around the tumor rim, which expressed CD31 and von Willebrand factor, indicating that the transplanted cells had differentiated into endothelial-like cells. In a different study, neural and progenitor cells and MSCs labeled with lipophilic dye-coated superparamagnetic particles were transplanted into rats 1 week after 9L-gliosarcoma cell transplantation. Three-dimensional gradient echo and T2-weighted images revealed the dynamic migration of adult neural progenitor cells and MSCs, detected as a hypointense signal, toward the tumor mass and infiltrating tumor cells. Prussian blue staining showed that areas with grafted cells corresponded to areas with a hypointense signal *(39)*.

1.6. MR Tracking of Implanted Cells in Other Organs

So far, we have described the use of MR imaging in the CNS. However, several successful applications for MR tracking can be found in other organs, such as heart, liver, kidney, and pancreatic islets. Iron fluorophore particles were used for labeling swine MSCs. Labeled MSCs were injected percutaneously into normal and freshly infarcted swine myocardium. Fluorophore particles provided sufficient MRI contrast for the detection of transplanted cells in the beating heart on a conventional cardiac MR scanner *(41)*. MSCs are potentially important vectors for cell and gene therapy in the liver or kidneys. MSCs labeled with SPIO and a transfection agent were injected into the renal artery trunk or portal vein to visualize SPIO-labeled MSCs within the kidney and liver parenchyma. However, the follow-up was limited to 7 or 12 days, therefore the fate of the cells and the SPIO label over the long term is not known *(42)*.

The contrast agent Resovist (a SPIO-type nanoparticle) was used as a marker of isolated pancreatic islets for MRI. Following transplantation into the liver of a rat, the labeled pancreatic islets could be easily detected as hypointense regions in the liver on T2*-weighted MR images for up to 22 weeks following transplantation. In addition, this method can be used for monitoring islet rejection in allogenic transplantations due to a decrease in the intensity of the hypointense regions during rejection *(43, 44)*.

2. Materials

2.1. Cell Culturing

1. Dulbecco's Modified Eagle's Medium (DMEM) – high glucose (4.5g/l) with L-Glutamin (PAA, Pasching, Austria) supplemented with 10% fetal bovine serum (FBS, PAA, Pasching, Austria) and with 0.2% primocin (Amaxa, Gaithersburg, MD, USA). Other antibiotics can be used instead of primocin:

100units/ml penicillin and 0.1 mg/ml streptomycin or gentamicin (50 μl/ml).

2. Phosphate-buffered saline (PBS) pH 7.2.

3. Solution of trypsin (0.5%) and ethylenediamine tetraacetic acid (EDTA) (0.2 g/l) from Gibco/Invitrogen (Carlsbad, CA, USA).

4. Solution of paraformaldehyde (4% in phosphate buffer).

5. Cell culture plasticware: centrifuge tubes, pipettes, Petri dishes (Technoplastic Products (TPP), Trasadingen, Switzerland), and tissue culture flasks (NUNC, Roskilde, Denmark).

6. Syringes (10 ml, Chirana, Slovak Republic), needles (18G, Becton Dickinson, Heidelberg, Germany).

2.2. Cell Labeling

1. Adherent cell culture, 40–50% confluent.

2. Solution of SPIO nanoparticles (Endorem, Guerbet Roissy, France). Stock solution contains 11.24 mg of iron per 1ml of Endorem®.

2.3. Prussian Blue Staining

1. Hydrochloric acid (0.5% HCl).

2. One gram of potassium ferrocyanide ($K_4[Fe(CN)_6] \cdot 3H_2O$; Sigma, St. Louis, USA) is diluted in 100 ml of 0.5% HCl 70% Ethyl Alcohol (EtOH).

3. Aluminum Sulfate (5% $Al_2(SO_4)_3$; Sigma, St. Louis, USA).

4. Nuclear fast red (Sigma, St. Louis, USA) diluted in 5% aluminum sulfate.

2.4. In Vitro MR Tracking

1. Set of Eppendorf tubes 1.6 ml (CPL Company, San Diego, CA, USA).

2. Cell suspension (2 million cells) labeled with SPIO nanoparticles and fixed in 4% paraformaldehyde.

3. Cell suspension (2 million cells) unlabeled and fixed in 4% paraformaldehyde (serves as control).

4. Gelatin (Sigma, St. Louis, USA)

2.5. Cell Implantation and In Vivo MR Tracking

1. Suspension of cells labeled with SPIO.

2. Animal with experimental model of injury or disorder. Since this protocol deals with MR tracking in living recipients, describing the induction of different models of injury is beyond the scope of this protocol.

3. MR equipment. The procedure may differ depending on the type (hole dimension, field strength, RF coil used, etc.) of the MRI imager. Procedures and listed sequence parameters are suitable for a standard horizontal bore imager 4.7 T.

3. Methods

MR tracking serves for studying how certain lesions target cell migration, at what speed the cells migrate, and for how long they persist in the target organ. High contrast effects on MR images are easily detected within an experimental time frame of about 1–2 h per animal, which is ideal for short and repetitive in vivo MRI. However, there are some disadvantages related to MR tracking in a living organism. The disadvantages include dilution of the contrast agent during cell division and a lack of information about cell viability or function. In addition, in lesioned tissue, hemorrhage products give rise to Prussian blue-positive deposits that are difficult to distinguish from iron-containing nanoparticles *(45)*. The hemorrhage degradation products may also be partially localized to macrophages because macrophages constitute the major cellular pathway for the redistribution of iron in mammals. Furthermore, hemorrhage contributes to T2-weighted hypointensity, thus interfering with the detection of labeled cells and complicating the interpretation of MR images. Therefore, it is important to determine whether cell labels remain colocalized with cell transplants, especially under pathological conditions.

With proper attention to the limitations described above, labeling cells with superparamagnetic agents would enable us to deliver the cells and immediately verify whether the cells have populated the target organ. Moreover, it will allow us to follow the migration of such cells when transplanted into humans, establish the optimal number of transplanted cells, define therapeutic windows and monitor cell growth and possible side effects, e.g., malignancies.

The described protocols are prepared for the isolation, expansion, and labeling of rat mesenchymal stem cells and their tracking in rat brain. However, these protocols can be modified to label different adherent cells with SPIO nanoparticles that can be kept in culture for 48–72 h. Similarly, other organs can be imaged as well.

3.1. Cell Culturing

Isolation and expansion of rat mesenchymal stem cells. Rat bone marrow mesenchymal stem cells (rMSCs) are obtained from the tibia and femur of 4-week-old rats. The ends of the bones are cut and the marrow is extruded with Dulbecco's Modified Eagle Medium (DMEM) by using a needle and a syringe. Marrow cells are plated in a 75-cm^2 tissue culture flask (TPP) in DMEM containing 10% fetal bovine serum (FBS) and 0.2% Primocin (Amaxa). After 24 h, the nonadherent cells are removed by replacing the medium. The medium is replaced every 2–3 days as the cells grow to confluence. The cells are lifted by incubation with 0.5% trypsin (Gibco/Invitrogen). The first passage after plating is usually taken as P0. Cells can be expanded up to passage 5 (P5) though the efficiency of cell labeling decreases with the number of passages.

3.2. Cell Labeling

A SPIO nanoparticle suspension (10 µl per 1 ml of culture medium corresponding to a final concentration of 112.4 µg of iron per 1 ml of culture medium) is added to a culture of rat MSCs or any adherent cells (*see* **Note 1**). After 72h the contrast agent is washed out, and the cells are harvested and used for either in vitro MR tracking or transplanted into animal models of disease or injury.

3.3. In Vitro MR Tracking

To verify whether the cells are properly labeled, cell phantoms should be prepared and measured by MRI in vitro.

1. *Preparation of phantoms containing suspensions of labeled cells.* To avoid deposition of the cells at the bottom of the test vials, the cells should be suspended in 4% gelatin. From a fixed cell suspension containing 2 million SPIO-labeled cells, several 1 mL samples can be prepared containing 100,000–1,000,000 cells/mL each. This range of concentrations ensures roughly 1–10 cells in one image voxel at a standard image geometry setting. Fixed cell suspensions in PBS containing the desired number of cells are placed into an Eppendorf tube in half of the final volume (i.e., 0.5 mL). Warm 8% gelatin (0.5 ml) is poured into the tube, mixed with the cells and quickly cooled on ice (*see* **Note 2**). The same procedure should be repeated with the unlabeled cells that serve as a control.

2. *Magnetic resonance imaging of cell phantoms.* For in vitro measurements we use standard sequences that are used for in vivo measurements, i.e., a T2-weighted turbospin echo sequence with the following parameters: TR = 2,000ms, effective echo time TE = 42.5ms, turbo factor = 4, number of acquisitions AC = 16, matrix 256 × 256, and slice thickness 0.5–1mm. The field of view is chosen according to the size of the samples. For T2*-weighted images, a gradient echo sequence with the same geometry can be used (*see* **Note 3**). Parameters of the sequence are TR = 80ms, TE = 5ms, and AC = 32.

3.4. In Vivo MRI

Superparamagnetic particles have a substantial impact on both T2 and T2* relaxation times. Therefore, both T2 (spin echo, turbospin echo) and T2*-weighted (gradient echo) sequences can be used. An implant with superparamagnetic particles will manifest itself as a hypointense area. Although T2*-weighted gradient echo is more sensitive to the presence of superparamagnetic particles, we strongly prefer a T2-weighted turbospin echo sequence, as it provides better anatomical images. The procedure may differ depending on the type (hole dimension, field strength, RF coil used, etc.) of MRI imager. The procedures and listed sequence parameters are suitable for a standard horizontal bore imager 4.7 T.

1. *Animal preparation.* The animal should be anesthetized throughout the entire measurement. Simple MRI is relatively short, therefore intubation of the animal is not necessary and passive inhalation of the anesthetic is sufficient for maintaining anesthesia. However, both passive and active (intubated and connected to an air pump) inhalation of a suitable anesthetic can be used. The anesthetized animal is placed and fixed into a heated holder dedicated for MR measurements. An airflow of 200–300 ml/min is maintained through a mouthpiece. The concentration of the inhalation anesthetic should be 1–2% depending on the breathing frequency of the animal. The breathing frequency is maintained at approximately one breath per second. Breathing is monitored throughout the entire experiment. A dedicated surface head coil is placed and fixed over the rat's head. (A suitable surface coil is better than whole body resonators because of its higher sensitivity.)

2. *Magnetic resonance imaging.* The holder with the animal is placed into the magnet bore. The RF coil must be connected and tuned. An automated procedure for shimming, setting of transmitter power, frequency, and receiver gain should be sufficient for MR imaging; however, if necessary, finely tuning may be done manually. Low resolution images are obtained in three orthogonal planes using a fast low angle gradient echo (FLASH) sequence (flip angle 30°, repetition time TR = 100ms, effective echo time TE = 5ms, matrix 128 × 128, slice thickness 3mm). The animal is repositioned if necessary. These images are used for positioning high resolution T2-weighted images. A sufficient number of slices to cover the entire volume of interest must be measured. Parameters of the sequence are TR = 2,000 ms, effective echo time TE = 42.5ms, turbo factor = 4, number of acquisitions AC = 16, field of view FOV = 3.5 cm, matrix 256 × 256, slice thickness 0.5–1mm, and slice separation 1mm. If the slice thickness is less than 1mm, it is advisable to measure two sets of interleaved images to avoid saturation of neighboring slices during the measurement. This measurement can be supplemented by a T2*-weighted gradient echo sequence with the same geometry; however, the procedure will be more time consuming. Parameters of the sequence are TR = 80ms, TE = 5ms, and AC = 32.

3.5. Staining for Iron

Fixed cells in culture wells or tissue slices are rinsed in distilled water, washed in alcohol (70%) for 2min, and twice rinsed in distilled water. Potassium ferrocyanide ($K_4[Fe(CN)_6] \cdot 3H_2O$) is applied for 30min to produce ferric ferrocyanide (Prussian blue). Cell nuclei are counterstained with nuclear fast red, or hematoxylin can be used (*see* **Note 4**).

4. Notes

1. *Cell labeling.* The concentration of iron can be decreased down to 15 µg of iron/ml of culture media. This positively influences cell viability; however, the labeling efficiency is lower (40–60%, depending on cell type and number of passage). Other SPIO nanoparticles, such as PCIO- or d-Mannose coated, can be used in lower concentrations (15 µg of iron/ml of culture media) without adversely affecting viability or labeling efficiency. The presence of nanoparticles in the cell culture medium can be decreased to 48h.

2. *Preparation of phantoms.* Tubes with cells suspended in gelatin can be vortexed before cooling so the cells are spread evenly.

3. *MRI of phantoms.* Cell samples should be imaged at room temperature, otherwise the gelatin partially melts and the samples become inhomogeneous.

4. We strongly recommend to simultaneously use another marker for detecting transplanted cells in histological slices. Particularly under pathological conditions, false Prussian blue positivity can result from deposits of iron in hemorrhages or in iron containing proteins (hemosiderin).

Acknowledgment

This work was supported by grants from the Academy of Sciences of the Czech Republic AV0Z50390703, KAN201110651, the Ministry of Education, Youth and Sports of the Czech Republic 1M0021620803, LC554, the National Grant Agency of the Czech Republic GACR 309/06/1594, the grant agency of the Ministry of Health NR8339-3, and the EC – FP6 project DiMI: LSHB-CT-2005–512146.

References

1. Emsley JG, Mitchell BD, Kempermann G, Macklis JD (2005). Adult neurogenesis and repair of the adult CNS with neural progenitors, precursors, and stem cells. *Prog Neurobiol* 75:321–341.

2. Fairless R, Barnett SC (2005). Olfactory ensheathing cells: their role in central nervous system repair. *Int J Biochem Cell Biol* 37:693–699.

3. McKay RD (2004). Stem cell biology and neurodegenerative disease. *Philos Trans R Soc Lond B Biol Sci* 359:851–856.

4. Newman MB, Davis CD, Borlongan CV, Emerich D, Sanberg PR (2004). Transplantation of human umbilical cord blood cells in the repair of CNS diseases. *Expert Opin Biol Ther* 4:121–130.

5. Park HC, Shim YS, Ha Y, Yoon SH, Park SR, Choi BH, Park HS (2005). Treatment of complete spinal cord injury patients by autologous bone marrow cell transplantation and administration of granulocyte-macrophage colony stimulating factor. *Tissue Eng* 11:913–922.

6. Pluchino S, Zanotti L, Deleidi M, Martino G (2005). Neural stem cells and their use as therapeutic tool in neurological disorders. *Brain Res Brain Res Rev* **48**:211–219.

7. Roitberg B (2004). Transplantation for stroke. *Neurol Res* **26**:256–264.

8. Uccelli A, Zappia E, Benvenuto F, Frassoni F, Mancardi G (2006). Stem cells in inflammatory demyelinating disorders: a dual role for immunosuppression and neuroprotection. *Expert Opin Biol Ther* **6**:17–22.

9. Zhao C, Fancy SP, Magy L, Urwin JE, Franklin RJ (2005). Stem cells, progenitors and myelin repair. *J Anat* **207**:251–258.

10. Mendonca Dias MH, Lauterbur PC (1986) Ferromagnetic particles as contrast agents for magnetic resonance imaging of liver and spleen. *Magn Reson Med* **3**:328–330.

11. Sykova E, Jendelova P (2005). Magnetic resonance tracking of implanted adult and embryonic stem cells in injured brain and spinal cord. *Ann N Y Acad Sci* **1049**:146–160.

12. Bonnemain A (1996). Superparamagnetic and blood pool agents. Spotlight on Clinical MRI. V.L.C., Maffliers, France, pp. 75–88.

13. Jendelová P, Herynek V, Urdzikova L, Glogarová K, Kroupová J, Bryja V, Andersson B, Burian M, Hájek M, Sykova E (2004). MR tracking of transplanted bone marrow and embryonic stem cells labeled by iron oxide nanoparticles in rat brain and spinal cord. *J Neurosci Res* **76**:232–243.

14. Kalish H, Arbab AS, Miller BR, Lewis BK, Zywicke HA, Bulte JW, Bryant LH, Jr., Frank JA (2003). Combination of transfection agents and magnetic resonance contrast agents for cellular imaging: relationship between relaxivities, electrostatic forces, and chemical composition. *Magn Reson Med* **50**:275–282.

15. Horák, D., Syková, E., Babic, M., Jendelová, P., and Hájek, M. (2006). Superparamagnetic Iron Oxide Nanoparticles with Modified Surface and Their Use as Probes for Stem Cell Labeling, PV 1006-120, Czech Republic.

16. Sykova E, Jendelova P (2007). In vivo tracking of stem cells in brain and spinal cord injury. *Prog Brain Res* **161**:367–83

17. Babic M, Horak D, Trchova M, Jendelova P, Glogarova K, Lesny P, Herynek V, Hajek M, Sykova E (2008). Poly(L-lysine)-modified iron oxide nanoparticles for stem cell labeling. *Bioconjug Chem* **19**:740–50.

18. Zhang J, Zhu J, Bu X, Cushion M, Kinane TB, Avraham H, Koziel H (2005). Cdc42 and RhoB activation are required for mannose receptor-mediated phagocytosis by human alveolar macrophages. *Mol Biol Cell* **16**:824–834.

19. Horak D, Babic M, Jendelova P, Herynek V, Trchova M, Pientka Z, Pollert E, Hajek M, Sykova E (2007). D-mannose-modified iron oxide nanoparticles for stem cell labeling. *Bioconjugate Chem* **18**(3):635–44.

20. Allen M, Bulte JW, Liepold L, Basu G, Zywicke HA, Frank JA, Young M, Douglas T (2005). Paramagnetic viral nanoparticles as potential high-relaxivity magnetic resonance contrast agents. *Magn Reson Med* **54**:807–812.

21. Bulte JW, Brooks RA, Moskowitz BM, Bryant LHJ, Frank JA (1999). Relaxometry and magnetometry of the MR contrast agent MION-46L. *Magn Reson Med* **42**:379–384.

22. Walczak P, Kedziorek DA, Gilad AA, Lin S, Bulte JW (2005). Instant MR labeling of stem cells using magnetoelectroporation. *Magn Reson Med* **54**:769–774.

23. Bulte JW, Hoekstra Y, Kamman RL, Magin RL, Webb AG, Briggs RW, Go KG, Hulstaert CE, Miltenyi S, The TH, et al. (1992). Specific MR imaging of human lymphocytes by monoclonal antibody-guided dextran-magnetite particles. *Magn Reson Med* **25**:148–157.

24. Jendelová P, Herynek V, Urdzíková L, Glogarová K, Rahmatová S, Fales I, Andersson B, Procházka P, Zame ník J, Eckschlager T, Kobylka P, Hájek M, Syková E (2005). MR tracking of human CD34+progenitor cells separated by means of immunomagnetic selection and transplanted into injured rat brain. *Cell Transplant.* **14**:173–182.

25. Jendelová P, Herynek V, De Croos J, Glogarová K, Andersson B, Hájek M, Syková E (2003). Imaging the fate of implanted bone marrow stromal cells labeled with superparamagnetic nanoparticles. *Magn Reson Med* **50**:767–776.

26. Hoehn M, Kustermann E, Blunk J, Wiedermann D, Trapp T, Focking M, Arnold H, Hescheler J, Fleischmann BK, Buhrle C (2002). Monitoring of implanted stem cell migration in vivo: A highly resolved in vivo magnetic resonance imaging investigation of experimental stroke in rat. *Proc Natl Acad Sci USA* **100**:1073–1078.

27. Modo M, Mellodew K, Cash D, Fraser SE, Meade TJ, Price J, Williams SC (2004). Mapping transplanted stem cell migration after a stroke: a serial, in vivo magnetic resonance imaging study. *Neuroimage* **21**:311–317.

28. Vanicky I, Urdzikova L, Saganova K, Cizkova D, Galik J (2001). A simple and reproducible model of spinal cord injury induced by epidural balloon inflation in the rat. *J Neurotrauma* **18**:1399–1407.

29. Urdzikova L, Jendelova P, Glogarova K, Burian M, Hajek M, Sykova E (2006). Transplantation

of bone marrow stem cells as well as mobilization by granulocyte – colony stimulating factor promote recovery after spinal cord injury in rat. *J Neurotrauma* **23**:1379–1391.

30. Lee IH, Bulte JW, Schweinhardt P, Douglas T, Trifunovski A, Hofstetter C, Olson L, Spenger C (2004). In vivo magnetic resonance tracking of olfactory ensheathing glia grafted into the rat spinal cord. *Exp Neurol* **187**:509–516.

31. Dunning MD, Kettunen MI, Ffrench Constant C, Franklin RJ, Brindle KM (2006). Magnetic resonance imaging of functional Schwann cell transplants labelled with magnetic microspheres. *Neuroimage* **31**:172–180.

32. Ben-Hur T, van Heeswijk RB, Einstein O, Aharonowiz M, Xue R, Frost EE, Mori S, Reubinoff BE, Bulte JW (2007). Serial in vivo MR tracking of magnetically labeled neural spheres transplanted in chronic EAE mice. *Magn Reson Med* **57**:164–171.

33. Dousset V, Ballarino L, Delalande C, Coussemacq M, Canioni P, Petry KG, Caille JM (1999). Comparison of ultrasmall particles of iron oxide (USPIO)-enhanced T2-weighted, conventional T2-weighted, and gadolinium-enhanced T1-weighted MR images in rats with experimental autoimmune encephalomyelitis. *AJNR Am J Neuroradiol* **20**:223–227.

34. Dousset V, Delalande C, Ballarino L, Quesson B, Seilhan D, Coussemacq M, Thiaudiere E, Brochet B, Canioni P, Caille JM (1999). In vivo macrophage activity imaging in the central nervous system detected by magnetic resonance. *Magn Reson Med* **41**:329–333.

35. Rausch M, Hiestand P, Baumann D, Cannet C, Rudin M (2003). MRI-based monitoring of inflammation and tissue damage in acute and chronic relapsing EAE. *Magn Reson Med* **50**:309–314.

36. Anderson SA, Shukaliak-Quandt J, Jordan EK, Arbab AS, Martin R, McFarland H, Frank JA (2004). Magnetic resonance imaging of labeled T-cells in a mouse model of multiple sclerosis. *Ann Neurol* **55**:654–659.

37. Rausch M, Hiestand P, Foster CA, Baumann DR, Cannet C, Rudin M (2004). Predictability of FTY720 efficacy in experimental autoimmune encephalomyelitis by in vivo macrophage tracking: clinical implications for ultrasmall superparamagnetic iron oxide-enhanced magnetic resonance imaging. *J Magn Reson Imaging* **20**:16–24.

38. Nakamizo A, Marini F, Amano T, Khan A, Studeny M, Gumin J, Chen J, Hentschel S, Vecil G, Dembinski J, Andreeff M, Lang FF (2005). Human bone marrow-derived mesenchymal stem cells in the treatment of gliomas. *Cancer Res* **65**:3307–3318.

39. Zhang Z, Jiang Q, Jiang F, Ding G, Zhang R, Wang L, Zhang L, Robin AM, Katakowski M, Chopp M (2004). In vivo magnetic resonance imaging tracks adult neural progenitor cell targeting of brain tumor. *Neuroimage* **23**:281–287.

40. Anderson SA, Glod J, Arbab AS, Noel M, Ashari P, Fine HA, Frank JA (2005). Noninvasive MR imaging of magnetically labeled stem cells to directly identify neovasculature in a glioma model. *Blood* **105**:420–425.

41. Hill JM, Dick AJ, Raman VK, Thompson RB, Yu ZX, Hinds KA, Pessanha BS, Guttman MA, Varney TR, Martin BJ, Dunbar CE, McVeigh ER, Lederman RJ (2003). Serial cardiac magnetic resonance imaging of injected mesenchymal stem cells. *Circulation* **108**:1009–1014.

42. Bos C, Delmas Y, Desmouliere A, Solanilla A, Hauger O, Grosset C, Dubus I, Ivanovic Z, Rosenbaum J, Charbord P, Combe C, Bulte JW, Moonen CT, Ripoche J, Grenier N (2004). In vivo MR imaging of intravascularly injected magnetically labeled mesenchymal stem cells in rat kidney and liver. *Radiology* **233**:781–789.

43. Jirak D, Kriz J, Herynek V, Andersson B, Girman P, Burian M, Saudek F, Hajek M (2004). MRI of transplanted pancreatic islets. *Magn Reson Med* **52**:1228–1233.

44. Kriz J, Jirak D, Girman P, Berkova Z, Zacharovova K, Honsova E, Lodererova A, Hajek M, Saudek F (2005). Magnetic resonance imaging of pancreatic islets in tolerance and rejection. *Transplantation* **80**:1596–1603.

Chapter 15

Identifying Neural Progenitor Cells in the Adult Brain

Stephen Kelly, Maeve Caldwell, Matthew P. Keasey, Jessica A. Cooke, and James B. Uney

Summary

There is incontrovertible evidence that neural progenitor cells (NPC) are found in the adult brain. The ability to identify and track NPC in the adult brain is of considerable importance if the properties of these cells are to be harnessed as potential therapies for degenerative brain disorders. The most commonly used approach of identifying these NPC in experimental studies, bromodeoxyuridine (BrdU) labelling, is outlined in this chapter. Immunohistochemical protocols for detecting endogenous and exogenous (introduced via transplantation) NPC in fresh-frozen and paraffin wax embedded brain tissue are described. Advice on how to label these NPC is also offered and multi-label fluorescence immunochemical staining approaches to determine the differentiation fate of NPC are described.

Key words: Bromodeoxyuridine, Neural progenitor cells, Immunohistochemistry, Immunofluorescence, Brain, Neurogenesis, Gliogenesis, Fresh-frozen, Paraffin embedded

1. Introduction

This chapter highlights the most common approaches to identify progenitor and stem cells in the adult brain. The techniques include the identification of endogenous neural progenitor cells (NPC) (sometimes referred to in the literature as multi-potent stem cells or neural precursor cells) and exogenous progenitor/stem cells that have been introduced into the brain (e.g. via transplantation).

The production of new neurons (neurogenesis) and glial cells (gliogenesis) is ongoing throughout adult life *(1)*. In mammals there are two major areas in the brain where new cells (referred to here as NPC) are produced. These are the sub-ventricular zone

Neil J. Scolding and David Gordon (eds.), *Methods in Molecular Biology, Neural Cell Transplantation vol. 549*
DOI: 10.1007/978-1-60327-931-4_15, © Humana Press, a part of Springer Science+Business Media, LLC 2009

of the lateral ventricles (SVZ) and the sub-granular zone of the dentate gyrus (SGZ). New cells from the SVZ migrate to the olfactory bulb (via the rostral migratory stream, RMS) (*2*) while new cells from the SGZ integrate into the local hippocampal network (*3*).

Exogenous NPC can be delivered to the brain via several different routes, including direct injection (transplant) or indirectly, such as, via tail-vein infusion. The ability to locate these cells after delivery is an essential part of any study of transplanted NPC. Locating same species NPC transplant offers several technical challenges whereas xenotransplant (e.g. human NPC into mouse) can often prove simpler.

2. Materials

2.1. Preparing Brain Tissue (Frozen and Paraffin Wax Embedded)

2.1.1. Common Items

1. Phosphate buffered saline (PBS, 1×), pH 7.4.
2. Fresh paraformaldehyde (PFA, 4% w/v in PBS, 1×), pH 7.4 (Fisher Scientific, UK).
3. Dry ice.
4. Isopentane (aka, methylbutane or 2-methylbutane) (Sigma-Aldrich, UK).
5. Sucrose/PFA (20% sucrose w/v in PFA 4% w/v) (Melford, UK).
6. OCT embedding compound (Raymond Lamb, UK).
7. Cryostat to cut sections at 15–40 µm.
8. Frozen section storage media (*see* below).

Should be used at pH 7.4	500 ml	1 l	2 l
Ethylene glycol	150 ml	300 ml	600 ml
Glycerol	100 ml	200 ml	400 ml
0.2 M Na_2HPO_4 stock	101 ml	202 ml	404 ml
0.2 M NaH_2PO_4·stock	24 ml	48 ml	96 ml
dH_2O	125 ml	250 ml	500 ml

9. Serial ethanol concentrations (50, 70, 90, 95 made up in distilled water and 100%) (Fisher Scientific, UK).
10. Xylene solutions (1:1 with 100% ethanol, 100% xylene) (Cellpath, UK).

11. Paraffin wax (liquid form in heating chamber above 60°C) (Raymond Lamb, UK).

12. Embedding cassettes and moulds (Raymond Lamb, UK).

2.2. BrdU Labelling (In Vivo and In Vitro)

2.2.1. In Vivo

1. BrdU, concentrations used vary by system and model being studied, typically a range of 10–200 mg/kg dissolved in sterile saline (NaCl) are used for animal injections. Make sure the volume delivered to the rat/mouse is around 0.5–1.0 ml (Sigma-Aldrich, UK).

2. Sterile syringe and needle (try to use as small a gauge and possible <25) (*see* **Note 1**).

2.2.2. In Vitro

1. BrdU doses historically range from 1 to 10 μM (*see* **Note 2**).

2.3. Immunohisto-chemistry (Chromogen-Based and Fluorescence)

1. Superfrost ++ glass slides (*for on-slide staining*) (VWR International, UK) OR a tissue culture plate, 12- or 24-well (Nunc, Denmark) (*for free-floating section staining*).

2. PBS (1×, pH 7.4).

3. Triton X100 (0.05–0.5%, in PBS 1×) (Sigma-Aldrich, UK).

4. Blocking solution (5–10% normal serum made up in PBS) (VectorLabs, CA, USA) (*the blocker varies with the source of the antibodies being used- details in Methods section*).

5. Primary antibodies.

6. Secondary antibodies.

7. Glass coverslips (*size depends on dimensions of stained sample*).

8. Small paint brush.

9. Large weigh boats.

10. Hydrogen peroxide (H_2O_2, 1–3% solution made up in ddH_2O) (Sigma-Aldrich, UK).

11. Avidin-Biotin Complex (ABC) *Elite* kit (VectorLabs, CA, USA).

12. Chromogen visualisation kit – Diaminobenzidine (DAB) or NovaRed (VectorLabs, CA, USA).

13. Haematoxylin for counterstain (VectorLabs, CA, USA).

14. Serial ethanol concentrations (70, 90, 95 made up in distilled water and 100%).[1]

15. Xylene solutions (1:1 with 100% ethanol, 100% xylene).

16. DPX mounting medium.

---

[1]Denotes optional – these can be replaced by allowing sections to air dry.

17. Fluorescence-conjugated secondary antibody (typically Cy-2, Cy-3, or Cy-5-conjugated AffiniPure IgG (H + L) (Jackson ImmunoResearch laboratories, Inc., USA).

18. Rat monoclonal anti-BrdU from Oxford Biotechnology, UK, cat# OBT0030CX.

19. 4′,6-diamidino-2-phenylindole (DAPI) 1:25,000 in PBS (1×) (Invitrogen, UK).

20. Vectashield mounting medium (which may contain DAPI or propidium iodide) (VectorLabs, CA, USA).

21. Nail varnish (Clear, to seal coverslip).

3. Methods

Pre-transplant NPC and those in the brain can be labelled by addition of ^{3}[H]-thymidine which would substitute for thymidine during cell division. This technique is often ignored nowadays in favour of 5-Bromo-2′-deoxyuridine (BrdU) labelling. BrdU is a thymidine analogue and is incorporated into the DNA of dividing cells during S-phase. It can be delivered by intraperitoneal injection into the animal or added to the culture media of NPC. BrdU has limitations (at high concentrations it can be toxic and can influence cell differentiation) but it remains a popular way of detecting dividing cells in the brain (and labelling pre-transplant NPC). There are two major reasons for this. First, delivery is relatively easy and can be adjusted simply. Second, it can be detected immunohistochemically. This means that once these cells have been identified as BrdU positive they can be targeted with other markers of cell lineage to determine the phenotypic fate of the NPC (or indeed whether it was an NPC to begin with). **Table 1** highlights many of the common antibodies used today as phenotypic markers in NPC studies. Two antibodies that can be used to directly detect dividing cells in the brain, Ki67 and proliferating cell nuclei antibody (PCNA), are also highlighted there.

Detecting NPC delivered into the host brain can be a relatively straightforward affair when cells from one species are infused into another. Probably the most common occurrence of this is when human NPC are delivered into rodent brain. Several good anti-human-cell-specific antibodies are available commercially for this approach. These in turn can be combined with phenotype markers to determine transplanted NPC fate. Identifying NPC from the same species (i.e. mouse NPC into mouse) can be somewhat more problematic. In this instance, it is common for the NPC being transplanted to be labelled in vitro prior to delivery. BrdU and ^{3}[H]-thymidine have been used for this approach with some

Table 1
Commonly used antibodies to identify proliferating cells and their phenotype

Cell type	Antibody	Comments	Reference
Proliferating	BrdU	Incorporates into proliferating cells DNA (S-phase)	*(4)*
	Ki67	Found in all active phases of the cell cycle but is absent in the resting phase (G0)	*(5)*
	PCNA	Proliferating cell nuclei antibody, expressed in the nuclei of actively dividing cells (S-phase)	*(6)*
Neural progenitors	Nestin	Cytoskeletal protein expressed in the axons of neurons and NPCs	*(7)*
	CD133	Glyocprotein expressed on the cell surface of NPC and glial stem cells	*(8)*
	Nucleostemin	Nuclear protein found in NPC	*(9)*
	Sox1	Early neuroectodermal marker. Expressed by NPCs	*(10)*
	Sox2	Earliest neuroectodermal marker to be expressed, maintains neural progenitor characteristics	*(10)*
Neurons	β III tubulin	Type 111 intermediate filament protein	*(11)*
	NeuN	Mature neuronal nuclear protein	*(12)*
	MAP2	Microtubule associated protein	*(13)*
	Synapsin	Neuronal phosphoproteins associated with cytoplasmic surface of synaptic vesicles	*(14)*
Human	Human nuclei	Labels human cell nucleus	*(15)*

success. Transduction (with viral vectors) and transfection of these pre-transplant NPC with reporter (marker) genes (such as EGFP) can also be used. There are limitations with each of these approaches. BrdU can have toxic effects and, like some viral vector and transfection paradigms, if the NPC continue to divide the label can become dilute. Moreover, there is some evidence that expression of some reporter transgenes in these systems can be suppressed, negating their utility.

The development of more sophisticated viral vectors (which enable long-term, NPC-promoter-specific reporter gene expression) is a recent way of identifying these cells (endogenous and exogenous). Moreover, transgenic mice which express EGFP under the control of the largely NPC-promoter-specific Nestin have also been generated to investigate the fate of these cells in vivo. These techniques require highly specialised training and are not available to all NPC researchers and will not be discussed further in this chapter.

3.1. Collecting and Processing Brains for Frozen and Wax Sectioning

3.1.1. Fixation

1. Terminally anaesthetise or kill the animal and immediately transcardially perfuse with chilled PBS (1×) followed by chilled PFA (4%). A peristaltic pump is the best way to do this. For standard sized rats (250–450 g) perfuse at least 150 ml of PBS and 250 ml of PFA. For standard size mice (22–40 g) perfuse at least 40 ml of PBS and 50 ml of PFA. Adjust volumes accordingly for larger/smaller rodents.

2. Remove the head and post-fix in PFA (4%) overnight at 4°C.

3. Carefully remove the brain and trim as appropriate for freezing or paraffin wax-embedding.

3.1.2. Paraffin Wax Processing and Sectioning

1. Although an entire mouse or rat brain can be treated for wax embedding if a vacuum embedder is used we cut our brains into blocks containing our regions of interest (e.g. sub-ventricular zone, dentate gyrus, olfactory bulb). *If no embedder is available this process can be carried-out manually on these smaller blocks of tissue* (*see* **Note 3** for details).

2. Using a brain matrix cut the brain into the appropriate blocks containing the area of brain you want to stain. We typically cut coronal blocks of around 4–6 mm.

3. Place these blocks into tissue processing cassettes and put back into PFA until they are ready to be processed.

4. In our case we use the standard processing and embedding cycle used in our hospital histopathology labs. In brief, the brains are washed in running water and then inserted into the chamber of the processing machine. Under vacuum they are then run through graded ethanols and xylenes before being impregnated with paraffin wax.

5. The wax impregnated blocks are then put in molten wax in a mould and placed on a cold-plate until the wax hardens (any cold surface can be used for this step).

6. The wax blocks can then be sectioned on a standard microtome. We collect sections of 7-μm thickness.

7. These sections are cut in ribbons (ideally 9–12 sections) and placed delicately on the surface of the water in a bath set at

~45°C until the sections flatten out completely. The wax between the sections is broken using fine-tipped forceps and the sections collected onto Superfrost ++ glass slides.

8. Excess water should be removed from under the sections by tapping the end of the slide on the bench.

9. Put the slides into a staining rack and place them on a hot-plate set at no more than 45°C until dry. Once dry store the sections in a dust-proof container at room temperature until ready for staining.

3.1.3. Frozen Brain Processing and Sectioning

1. To reduce ice-crystal artefact on brain slices the brain must be cryo-protected. Place the brain into at least 40 ml of chilled sucrose/PFA (20% sucrose w/v in PFA 4% w/v) and store at 4°C. The brain will initially float in this mixture and as the sucrose replaces water in the tissue it will drop to the bottom of the container. When the brain sinks (normally 1–2 days), proceed.

2. Pour some isopentane (100–500 ml, depending on number of brains to freeze) into a metal beaker and place on dry-ice. Chill the liquid to ~–42°C and immerse the brain in there for 5–10 min. The brain can now be mounted for sectioning (described below) or stored (*see* **Note 4**).

3. Some cryostats have rapid freezing areas for brain mounting. We prefer to mount the brain using powdered dry-ice. Mounting the brain should be done quickly and precisely. Place the cryostat chuck into the dry-ice (for 20 s) to chill it down and cover the surface with OCT mounting media. As the OCT solution turns white (freezes) place your brain in it in the orientation you desire and keep it chilled on dry-ice. The OCT should set around the area of brain submerged in it.

4. The chuck (with brain attached) should now be placed into the cryostat and the temperature allowed to equilibrate (the brain will be significantly colder than the air in the cryostat) before sectioning should commence (to avoid fracture of the tissue).

5. Sections can now be collected directly onto Superfrost + + glass slides or placed into a tissue culture dish containing storage media (remains liquid at –20°C). For sections placed directly onto glass slides we cut at 20 μm, for free-floating sections we cut at 40 μm.

6. Store sections appropriately until required for staining.

3.2. Chromogen-Based BrdU Immunohisto-chemistry

All steps are carried out at room temperature unless stated otherwise.

3.2.1. Pre-Treatment of Wax Sections

1. Remove wax from tissue by placing slides into a rack and immersing in fresh xylene – 2 × 15 min.

2. Put slides into a 1:1 mixture of xylene and ethanol (100%) 1 × 15 min.

3. Rehydrate sections through graded ethanols – 100% (2 × 5 min), 95%, 70% and 50% (1 × 5 min).

4. Wash slides in running tap water for 5 min.

5. Wash slides in PBS – 2 × 5 min.

6. Wash slides in PBS^{++} (which is PBS with 0.3% TX100 v/v) – 3 × 5 min.

3.2.2. Pre-Treatment of Frozen Sections

1. Remove frozen sections from freezer and allow them to equilibrate to room temperature.

2. Wash with PBS – 2 × 5 min.

3. Wash with PBS^{++} (which is PBS with 0.3% TX100 v/v) – 3 × 5 min.

3.2.3. Common Pre-Treatment Steps

1. To quench endogenous peroxidase activity, immerse the sections in hydrogen peroxide (3% in PBS^{++}) for 20 min. On frozen sections, particularly free-floating sections, you should see formation of small bubbles as the quenching takes place.

2. Wash with PBS^{++} 3 × 5 min (making sure that all of the bubbles have been removed, wash again until this is achieved).

3. Wash with sodium chloride (0.9%) – 3 × 5 min.

4. Immerse sections in 2 M hydrochloric acid (HCl) for 10 min.

5. Immerse sections in 3 M HCl for 30 min at 37°C (*see* **Note 5** for further details).

6. Wash sections thoroughly with PBS^{++} – 3 × 5 min.

3.2.4. Blocking and Antibody Steps

1. Block sections with normal goat serum (10% in PBS) – 1 h.

2. After removing blocker, add primary antibody against BrdU and incubate overnight at 4°C. End user should determine antibody dilution to be used depending upon their sample conditions (*see* **Note 6**).

3. Retrieve primary antibody for re-use (generally this can be used a few times – make sure to test this on non-essential samples first).

4. Wash sections with PBS – 3 × 5 min.

5. Add secondary antibody, anti-rat IgG (1:200 dilution in PBS) for 1 h. *After 40 min prepare the ABC solution for use later.*

6. Remove secondary antibody and wash with PBS – 3 × 5 min.

7. Incubate with ABC solution for 1 h.

8. Wash with PBS – 3 × 5 min.

9. Add chromogen of choice according to manufacturers' instructions. We use either diaminobenzidene (DAB) or NovaRed kits from VectorLabs. In short, the solutions are mixed together and the sections incubated for up to 8 min – until positive staining can be seen under a microscope (*see* **Note** 7).

10. Wash sections thoroughly with water to stop peroxidase reaction (1 × 5 min).

11. At this point the sections need to be dehydrated (or mounted onto slides and dehydrated if free-floating section have been used – see below for mounting instructions)

12. To dehydrate the sections they can simply be left in a slide rack to dry at room temperature or they can be run through graded ethanols and xylene (i.e. repeat **steps 3, 2** and **1** of this protocol – in that order) and coverslipped using DPX.

3.2.5. Mounting Free-Floating Sections onto Slides

1. Lift stained sections using a fine paint brush and place into a suitable receptacle (we use a large plastic weigh boat on a black card surface – to increase contrast) filled with PBS. To reduce surface tension put a tiny drop of Triton-X100 into the PBS (~2 μl).

2. Dip your glass slide into the PBS and use your paint brush to drag your section across the slide and flatten it out. Repeat as necessary and allow the sections to air dry.

3. Once dry, the sections should be dipped into xylene and coverslipped using DPX.

3.2.6. Counterstaining

To put the BrdU labelled cells into context it is often useful to counterstain your sections. This should be done prior to the dehydration step for sections already on glass slides and after dehydration of free-floating sections that have been mounted onto slides. Several counterstains are available; we use haematoxylin. The idea of a counterstain is to very lightly label the cells in your sections to help identify structures without drowning out the immunoreactive cells (in this case BrdU positive cells) (*see* **Note 8**).

1. As the haematoxylin is water based, the sections must be hydrated. Those on glass slides can be used immediately, those dried following mounting onto slides need to be rehydrated through graded ethanols back into water (**steps 2–4** of this protocol).

2. Put sections into haematoxylin for 20 s, immediately wash in running tap water and observe staining under a microscope. Continue this process until a good, light stain is observed (*see* **Note 9**).

3. When you are happy with the stain (it should be slightly darker than you want it to be in the end) wash the sections in running tap water for 2 min.

4. If the staining is a bit strong you can differentiate it by dipping into hydrochloric acid (1%) in methanol. Each dip will reduce the level of haematoxylin stain (*see* **Note 10**).

5. Once a suitable counterstain has been achieved, dehydrate the sections through graded ethanols and xylene as described in **steps 3, 2** and **1**.

6. Coverslip sections using DPX.

7. When DPX is dry, sections can be viewed using a light microscope.

3.3. Fluorescence-Based BrdU Immuno-histochemistry

We rarely use fluorescence labelling for wax sections, it does, however, have the advantage of requiring fewer steps and allows for cells to be labelled with more than one antibody.

3.3.1. Pre-Treatment of Wax Sections

1. Remove wax from tissue by placing slides into a rack and immersing in fresh xylene – 2 × 15 min.

2. Put slides into a 1:1 mixture of xylene and ethanol (100%) 1 × 15 min.

3. Rehydrate sections through graded ethanols – 100% (2 × 5 min), 95%, 70% and 50% (1 × 5 min).

4. Wash slides in running tap water for 5 min.

5. Wash slides in PBS – 2 × 5 min.

6. Wash slides in PBS^{++} (which is PBS with 0.3% TX100 v/v) – 3 × 5 min.

3.3.2. Pre-Treatment of Frozen Sections

1. Remove frozen sections from freezer and allow them to equilibrate to room temperature.

2. Wash with PBS – 2 × 5 min.

3. Wash with PBS^{++} (which is PBS with 0.3% TX100 v/v) – 3 × 5 min.

3.3.3. Common Pre-Treatment Steps

1. Wash with sodium chloride (NaCl, 0.9%) – 3 × 5 min.

2. Immerse sections in 2 M hydrochloric acid (HCl) for 10 min.

3. Immerse sections in 3 M HCl for 30 min at 37^{0}C.

4. Wash sections thoroughly with PBS^{++} – 3 × 5 min.

3.3.4. Blocking and Antibody Steps

1. Block sections with normal donkey serum (5% in PBS) – 1 h.

2. After removing blocker, add primary antibody against BrdU (Rat monoclonal anti-BrdU from Abcam, cat# ab6326) and incubate overnight at 4°C. End user should determine antibody dilution to be used depending upon their sample conditions (*see* **Note 6**).

3. Retrieve primary antibody for re-use (generally this can be used a few times – make sure to test this on non-essential samples first).

4. Wash sections with PBS – 3 × 5 min.

5. Add secondary antibody, donkey anti-rat IgG with Cy3 conjugated (1:400 dilution in PBS) for 1 h.

6. Remove secondary antibody and wash with PBS – 3 × 5 min.

7. At this point cell nuclei can be labelled by washing with DAPI in PBS (1:25,000 dilution) for 5 min.

8. Remove DAPI and wash with PBS – 3 × 5 min.

9. At this point the sections should be kept "wet" and either mounted onto slides (if not already on slides) and coverslipped using Vectashield – if no further labels are to be added.

10. If you wish to probe with further antibodies (e.g. lineage markers – neuronal or glial) then immerse the sections in PFA for 20 min and follow the protocol in **Subheading 3.4, step 2**.

3.4. Multi-label Fluorescence Staining to Determine NPC Phenotype

As outlined above NPC loaded (in vivo or in vitro) with BrdU can be detected immunohistochemically. BrdU, however, is not an NPC specific marker (neither are Ki67 nor PCNA which can also be used). They will detect all dividing cells. This type of labelling is adequate where the experiment is designed to show an overall increase or decrease in cell proliferation in the brain, but is inadequate to determine if NPC alone are altered. To do this, one must use multi-label fluorescence to identify the phenotype of BrdU labelled cells.

We typically use 40-µm thick, free-floating frozen sections for this approach, although slide mounted frozen and wax sections can also be used. The protocol below describes our standard triple-label approach for BrdU (or Ki67) with β III tubulin (Tuj1 antibody) and GFAP.

3.4.1. Pre-Treatment of Frozen Sections

1. Remove frozen sections from freezer and allow them to equilibrate to room temperature or if continuing with BrdU stained sections remove PFA.

2. Wash with PBS – 2 × 5 min.

3. Wash with PBS^{++} (which is PBS with 0.1% TX100 v/v) – 3 × 5 min.

3.4.2. Blocking and Antibody Steps

1. Block sections with normal donkey serum (5% in PBS) – 1 h.

2. After removing blocker, add primary antibodies against β III tubulin (we use mouse monoclonal anti-β III tubulin from Abcam.com, cat# ab14545) and GFAP (guinea pig polyclonal anti-GFAP from Advanced ImmunoChemicals, cat# 31223) incubate overnight at 4°C. End user should

determine the optimum antibody dilution to be used for their sample conditions (*see* **Note 11**).

3. Retrieve primary antibody for re-use (generally these can be used a few times – make sure to test this on non-essential samples first).

4. Wash sections with PBS – 3 × 5 min.

5. Add secondary antibodies, donkey anti-mouse IgG with Cy2 conjugated and donkey anti-guinea pig with Cy5 conjugated (each at 1:400 dilution in PBS) for 1 h.

6. Remove secondary antibody and wash with PBS – 3 × 5 min.

7. At this point, cell nuclei can be labelled by washing with DAPI in PBS (1:25,000 dilution) for 5 min.

8. Remove DAPI and wash with PBS – 3 × 5 min.

9. Keep the sections "wet" and mount onto slides.

10. Coverslip using Vectashield.

11. To prevent drying of the sections and loss of staining it is prudent to secure the coverslip over the sections using clear nail varnish.

12. Once dried the sections can be viewed under a fluorescence microscope with a tungsten light source and filters for green, red, far red and dapi. Do this for a very short time (seconds) just to check that the staining has worked.

13. If the staining has worked, the sections should be stored in a cool dark place until they can be viewed on a confocal microscope.

14. The confocal microscope must have excitation lasers to stimulate dapi (364 nm) for blue emission (454 nm), Cy2 (492 nm) for green emission (510 nm), Cy3 (550 nm) for red emission (570 nm) and Cy5 (650 nm) for far red emission (670 nm).

3.5. Detecting Transplanted NPC

As mentioned earlier in the text, generally speaking, there are two scenarios here. Xenotransplant and same species transplant. We commonly detect human NPC in rodent brain using the same protocols already outlined in **Subheading 3.2–3.4**, the only differences being the primary antibody and blocking sera used. To stain for human cells we use anti-human nuclei antibody (from Chemicon, USA) 1:50 dilution in PBS and normal horse serum for chromogen-based staining. In order to detect same species (i.e. mouse NPC in mouse brain) pre-transplant labelling is required. Depending upon the labelling techniques employed, these cells can then be detected as described in **Subheading 3.2–3.4** using antibodies directed at it, for example, BrdU.

4. Notes

1. Smaller diameter needles cause less damage during repeated injections. Sometimes injections are required several times a day or daily over several days. We have found that syringes with fixed needles designed for daily insulin delivery for diabetics work well.

2. A recent paper (*16*) suggests that 0.2 μM is a more effective dose.

3. Wax embedding by hand (for small volumes of tissue): (a) Cut PFA (4%) fixed tissue into small blocks and place into tissue embedding cassettes and immerse cassettes in the following solutions in order; (b) Running water for at least 30 min; (c) 70% Ethanol, 2 × 30 min; (d) 90% Ethanol, 2 × 30 min; (e) 100% Ethanol, 2 × 30 min; (f) 100% Ethanol/Xylene (1:1), 30 min; (g) Xylene, 2 × 30 min; (h) Molten paraffin wax (60°C), 1 h; (i) Molten paraffin wax (60°C), overnight; (j) At this point follow **steps 8–12** of **Subheading 3.1**.

4. To store the brain and avoid freezerburn, wrap it in Parafilm, place in a small sealable bag, put in a closable container (like a 50 ml Falcon tube); and store until needed at −80°C.

5. These steps may be lengthened or shortened depending upon the condition of the tissue being stained and the age of the HCl. Ideally "fuming" acid should be used. Moreover, the acid steps are often followed by immersing the sections in borate buffer. This is an optional step which we do not always use. We generally use generous PBS washes.

6. As reference for 7-μm wax sections we use 1:200 dilution (in PBS) for 40-μm frozen sections we use 1:250.

7. By stopping the reaction with water on a test section every minute or so to check staining level the precise incubation time for your sample can be determined for the remaining sections.

8. If your BrdU labelling is very faint counterstaining is best avoided.

9. This step is completely end-user dependent as differently processed tissues and different haematoxylin mixtures (including older solutions vs. fresh solutions) vary from lab to lab. It may be the case that you can increase the amount of time between each observation of the staining.

10. Take care after this treatment to get the sections back into running tap water swiftly to avoid excessive destaining. If too much stain is removed – repeat from **step 2** in Section 3.2.6.

11. As reference for 7-μm wax sections we use β-tubulin III at 1:600 and GFAP at 1:300 dilutions in PBS. For 40-μm frozen sections we use β-tubulin III at 1:1,000 and GFAP at 1:500 dilutions in PBS.

References

1. Gould, E. (2007). How widespread is adult neurogenesis in mammals? *Nat Rev Neurosci*, **8**: 481–8.

2. Curtis, M.A., et al. (2007). Human neuroblasts migrate to the olfactory bulb via a lateral ventricular extension. *Science*, **315**(5816): 1243–9.

3. Kempermann, G., et al. (2004). Milestones of neuronal development in the adult hippocampus. *Trends Neurosci*, **27**(8): 447–52.

4. Kuhn, H.G., et al. (1997). Epidermal growth factor and fibroblast growth factor-2 have different effects on neural progenitors in the adult rat brain. *J Neurosci*, **17**(15): 5820–9.

5. Komitova, M., et al. (2005). Enriched environment increases neural stem/progenitor cell proliferation and neurogenesis in the subventricular zone of stroke-lesioned adult rats. *Stroke*, **36**(6): 1278–82.

6. Redmond, L., S. Hockfield, and M.A. Morabito (1996). The divergent homeobox gene PBX1 is expressed in the postnatal subventricular zone and interneurons of the olfactory bulb. *J Neurosci*, **16**(9): 2972–82.

7. Lendahl, U., L.B. Zimmerman, and R.D. McKay (1990). CNS stem cells express a new class of intermediate filament protein. *Cell*, **60**(4): 585–95.

8. Uchida, N., et al. (2000). Direct isolation of human central nervous system stem cells. *Proc Natl Acad Sci USA*, **97**(26): 14720–5.

9. Maki, N., et al. (2007). Rapid accumulation of nucleostemin in nucleolus during newt regeneration. *Dev Dyn*, **236**(4): 941–50.

10. Graham, V., et al. (2003). SOX2 functions to maintain neural progenitor identity. *Neuron*, **39**(5): 749–65.

11. Svendsen, C.N., et al. (1998). A new method for the rapid and long term growth of human neural precursor cells. *J Neurosci Methods*, **85**(2): 141–52.

12. Mullen, R.J., C.R. Buck and A.M. Smith (1992). NeuN, a neuronal specific nuclear protein in vertebrates. *Development*, **116**(1): 201–11.

13. Diez-Guerra, F.J. and J. Avila (1993). MAP2 phosphorylation parallels dendrite arborization in hippocampal neurons in culture. *Neuroreport*, **4**(4): 419–22.

14. Rastaldi, M.P., et al. (2006). Glomerular podocytes contain neuron-like functional synaptic vesicles. *FASEB J*, **20**(7): 976–8.

15. Fricker, R.A., et al. (1999). Site-specific migration and neuronal differentiation of human neural progenitor cells after transplantation in the adult rat brain. *J Neurosci*, **19**(14): 5990–6005.

16. Caldwell, M.A., X. He, and C.N. Svendsen (2005). 5-Bromo-2′-deoxyuridine is selectively toxic to neuronal precursors in vitro. *Eur J Neurosci*, **22**(11): 2965–70.

Chapter 16

Immune Ablation Followed by Autologous Hematopoietic Stem Cell Transplantation for the Treatment of Poor Prognosis Multiple Sclerosis

Harold Atkins and Mark Freedman

Summary

Complete abrogation of the inflammatory response by high-dose cytotoxic therapy at an early stage of MS, when the nervous system has not yet sustained irreparable damage may be successful at preventing the inexorable progression. Immunological and hematological reconstitution follows abrogation through bone marrow transplantation. The issues are complex, and many factors, including baseline disability, the timing of this intervention, the intensity of the immune ablation, and depletion of lymphocytes from the graft, are all likely to influence treatment outcome. This article describes the immune ablation regimen for treatment of patients with poor prognosis MS, as performed in the Canadian MS-BMT study.

Key words: Multiple sclerosis, Bone marrow transplantation, Immune ablation

1. Introduction

Though not yet clearly proven, the damage caused by early inflammatory events in multiple sclerosis (MS) is thought to spearhead a neurodegenerative process that begins early but dominates in later stages of disease. Complete abrogation of the inflammatory response by high-dose cytotoxic therapy at an early stage of MS, when the nervous system has not yet sustained irreparable damage may be successful at preventing the inexorable progression. Preclinical reports demonstrating the beneficial effect of high-dose cytotoxic therapy followed by

Neil J. Scolding and David Gordon (eds.), *Methods in Molecular Biology, Neural Cell Transplantation vol. 549*
DOI: 10.1007/978-1-60327-931-4_16, © Humana Press, a part of Springer Science+Business Media, LLC 2009

pseudoautologous marrow transplantation in the treatment of experimental allergic encephalitis (EAE) appeared in the early 1990s (*1, 2*). The first clinical use of autologous hematopoietic stem cell transplantation (HSCT) for the treatment of patients with MS was reported soon thereafter (*3*). The role of HSCT in the treatment of MS has recently been reviewed (*4, 5*). As of 2006, reports to the European Bone Marrow Transplant Registry (*6*) (EBMTR) and the Center for International Blood and Marrow Transplant Research (*7*) (CIBMTR) document that more than 250 patients have received autologous stem cell transplants for the treatment of refractory MS.

Baseline disability, the intensity of the immune ablation, and depletion of lymphocytes from the graft all influence treatment outcome. Patients with a pretreatment Kurtzke Expanded Disability Status Score (EDSS) of 6.5 or more tend to have sustained disease progression following transplantation compared to those with a pre-transplant EDSS of 6.0 or less. High intensity conditioning regimens such as Cyclophosphamide with total body irradiation (*8, 9*) or Busulphan with Cyclophosphamide, as used in our study, provide better disease control than less aggressive regimens such as BEAM (*1*), however, the more intensive HSCT regimens also have greater acute toxicity. Adoptive transfer of aberrant immune responses has been documented in recipients of lymphocyte-replete hematopoietic stem cell (HSC) grafts; thus MS may persist if autoreactive lymphocytes are reintroduced into the patient via the HSC graft (*10, 11*). While, there was no difference in MS disease response following HSCT with lymphocyte-replete or lymphocyte-deplete grafts in one study (*12*), technical issues such as poor lymphocyte depletion using first generation stem cell selection technology and a low intensity HSCT regimen complicate its interpretation.

This article describes the immune ablation regimen for treatment of patients with poor prognosis MS, as performed in the Canadian MS-BMT study.

2. Patient Selection

Patients considered for this treatment must have a diagnosis of active multiple sclerosis made by a neurologist expert in the field, and demonstrate relapses or progression, with sustained accumulation of impairment. Their MRI brain scan must satisfy the MRI criteria of Paty or Fazekas for the diagnosis of multiple sclerosis. Evidence for current activity includes a deterioration in the EDSS score of 1 or more in the 18 months prior to enrollment if the EDSS score is <5.5, a deterioration in the EDSS score of 0.5 or more in the 18 months prior to enrollment if the

EDSS score is ≥5.5 or at least two significant relapses in the last year, or three significant relapses in the last 2 years. Patients must have an EDSS score from 3.0 to 6.0 with an EDSS Cerebellar Functional score ≥3 or EDSS Pyramidal Functional score ≥3. We feel that these patients have disabilities that are significant enough to impact their lifestyle and yet are likely to be a stage where the CNS lesions are due more to inflammation than to scarring. The patient should have a sufficiently poor prognosis (a high probability of progression according to the criteria of Weinshenker (13)) to warrant the risks associated with this therapy. Poor prognostic features include multiple relapses (e.g., ≥5 in the first 2 years of disease), or attaining a Kurtzke Functional System Score (FS) of at least 3 (or findings consistent with a FS of 3) affecting pyramidal/cerebellar subscores within 5 years of onset. Patients should have failed standard therapy for MS as shown by progression or continued relapses or worsening MRI after at least 1 year of therapy with Interferon-β, Glatiramer acetate, Mitoxantrone, or other immunosuppressive drug therapy.

Patients who have previously received a cytotoxic agent (Mitoxantrone, Cyclophosphamide, etc.) must have normal bone marrow morphology and cytogenetics before being considered for HSCT. Significant cardiac, renal, pulmonary, or hepatic dysfunction, active infections, or other medical problems that would appreciably increase the patient's risk of morbidity or mortality, are contraindications to HSCT.

3. Preparative Phase

A bone marrow is cryopreserved, to be used in case of persistent pancytopenia or immunodeficiency following the CD34-selected autologous HSCT. The collection of this autologous graft is performed as a day hospital surgery. An indwelling central venous catheter is placed at the same time to provide vascular access during HSCT (14, 15).

3.1. Autologous Bone Marrow Harvest

The patient is taken to the operating room and placed under general anesthesia. Once intubated and ventilated, the patient is rolled into the prone position on bolsters that support the trunk, head, and neck. The patient is prepped from the base of the buttocks to the mid-back. An 8-gauge bone marrow aspiration needle (Snarelock bone marrow biopsy needle, Angiotech) is inserted percutaneously in the posterior iliac crest and the trochar removed. A 60-ml syringe, pre-rinsed in a heparin solution (10,000 unit heparin per liter NS), is attached to the needle and 10–20 ml of marrow is aspirated using vigorous suction. The needle is removed and the marrow is collected in a marrow graft collection bag (Bone Marrow Collection Kit R4R2107, Baxter) containing 100 ml of ACD-A (2) per liter of bone marrow. The needle is rinsed in heparin solution and the trochar is replaced before re-use. The aspiration is repeated multiple times by two surgeons, one working on the right side and the other working on the left side until a total of 10–15 ml of bone marrow per kg of patient weight is collected. At the end of the procedure, the marrow is filtered through sequential 800–200-µm filters to remove coarse particles.

3.2. Quality Testing of the Bone Marrow Graft

An aliquot of the filtered marrow graft is tested for microbiological sterility. 0.5–1.0 ml of the graft is added to a standard blood culture sample vial and processed using standard techniques for blood cultures in the hospital microbiology lab.

Bone marrow mononuclear cells are enumerated using a Coulter counter. HSCs are enumerated by flow cytometry following staining with anti-CD45-FITC and anti-CD34-PE using the ISHAGE method (16). The bone marrow graft must contain a minimum of 1.5×10^8 marrow mononuclear cells/kg of recipient weight to be considered adequate, before the patient proceeds to the next stage of the clinical protocol.

3.3. Bone Marrow Graft Cryopreservation

The bone marrow graft is cryopreserved according to our stem cell laboratory's standard operating procedures. Briefly, the bone marrow buffy coat is isolated from the bone marrow graft following centrifugation on a Cobe 2291 Cell Processor. The bone marrow buffy coat is mixed with an equal volume of chilled cryoprotectant solution (3) and divided into two aliquots which are transferred to individual cryobags (Cryocyte Freezing Container, Baxter) for storage in the vapor phase of a liquid nitrogen container.

3.4. Vascular Access (3)

Following the marrow collection, while still under general anesthesia, the patient is rolled into the supine position. A central venous catheter is inserted according to the manufacturer's procedures. Briefly, the surgeon creates a subcutaneous tunnel by blunt dissection of the subcutaneous tissues between an incision on the anterior chest wall (exit site) and an ipsilateral subclavicular incision (entrance site). Using a Seldinger technique at the entrance site, a double-lumen central venous catheter is placed into the subclavian vein and advanced into the superior vena cava under fluoroscopic control. The catheter is tunneled from the entrance to the exit incision on the anterior chest wall and positioning the catheter's Dacron cuff just inside the exit tunnel. The incisions are sutured closed and the patient is awoken from anesthesia. A standard chest radiograph is used to confirm the position of the catheter tip and the absence of pneumothorax at the end of the procedure.

3.5. Postoperative Analgesia

Postoperative analgesia is accomplished with acetaminophen (650 mg) with codeine (30 mg po) every 4 h as required.

3.6. Central Venous Catheter Care

The central venous catheter is cared for according to hospital procedures, the exit site is cleaned with a chlorhexidine disinfectant and the proximal external portion of the catheter held down by a sterile gauze or occlusive dressing centered on the exit site. The dressing is changed on the day following insertion and then once weekly. Each lumen is flushed with 10 ml of a diluted heparin solution (1 ml of 100 units/ml heparin with 9 ml NS) after each use or once weekly when the lumen is not in use.

4. Autologous Stem Cell Mobilization and Collection (*See* Figs. 1 and 2)

Autologous HSC are collected and cryopreserved as a source of stem cells for the reconstitution of the bone marrow and the immune system prior to administering the immune ablative therapy. Chemotherapy and recombinant human granulocyte-colony stimulating factor (rhG-CSF) stimulate the mobilization of HSC into the circulation. The circulating white blood cells (WBC), now enriched in the patient's own HSC, can be collected by leukopheresis. The HSCs are purified away from contaminating immune cells in the graft by immunomagnetic selection of cells expressing CD34, a HSC expressed antigen. While hematopoietic growth factors alone can be used to mobilize HSC into the circulation, it has been associated with flares of the underlying autoimmune disease. Coadministration of steroids and cytotoxic chemotherapy prevent the cytokine-associated exacerbation of the MS.

4.1. High-Dose Cyclophosphamide Administration

The chemotherapy is administered in an inpatient hospital setting. High-dose cyclophosphamide is considered a highly emetogenic chemotherapy. The risk of developing chemotherapy-induced nausea is minimized using HT_3 antagonists and steroids. Hyperhydration, Foley catheter drainage of the urinary bladder, and

Baseline Evaluations ~2 months	Bone Marrow Harvest and Central Venous Catheter Insertion	Autologous Hematopoietic Stem Cell Mobilization and Peripheral Blood Collection	CD34 selection of autologous peripheral blood stem cell graft	Immune Ablation followed by Autologous Hematopoietic Stem Cell Transplantation	Recovery and ongoing evaluation of MS
~2 months	~1 day	~2 weeks		~1-3 months	Ongoing for >1 year

Fig. 1. Timeline for immune ablative therapy and autologous stem cell transplantation for patients with poor prognosis multiple sclerosis.

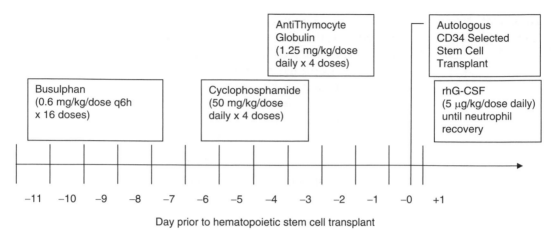

Day prior to hematopoietic stem cell transplant

Fig. 2. Timeline for administration of immune ablation regimen and hematopoietic stem cell transplantation.

Sodium 2-mercaptoethane sulfonate (MESNA) administration are used to minimize the risk of hemorrhagic cystitis, a complication of cyclophosphamide administrations that is of particular concern in MS, given the bladder dysfunction that often accompanies this illness.

4.1.1. Hyperhydration

Normal saline (NS) with 20 mEq KCl/l is run intravenously at 150 ml/h starting from the evening before chemotherapy and continuing for 24 h following completion of cyclophosphamide. A Foley catheter is inserted to drain the urinary bladder and monitor urine output during hyperhydration. Body weight and fluid balance (ins and outs) are monitored every 8 h during the period of hyperhydration. Furosemide (20 mg) is administered intravenously if the patient gains more than 1 kg in an 8-h period or has more than 1-l positive fluid balance.

4.1.2. Mesna

Mesna (600 mg/m² in 100-ml NS) is administered as an intravenous bolus over 30 min just prior to Cyclophosphamide. The bolus is followed by a continuous intravenous infusion of Mesna (4,800 mg/m² in 1,000 ml NS at 42 ml/h) which is continued for 24 h following completion of Cyclophosphamide.

4.1.3. Anti-Emetic Prophylaxis

Ondansetron (8 mg po or iv) or an equivalent HT_3 receptor blocker is administered with Dexamethasone (8 mg po or iv) 30 min prior to Cyclophosphamide and are then given every 12 h for six doses.

4.1.4. Cyclophosphamide

Cyclophosphamide (4.5 g/m² in 1,000 ml NS) is given as an intravenous dose of over 3 h. The body surface area is calculated using Mosteller formula (*17*) and the lesser of the patient's actual or ideal body weight.

4.2. rhG-CSF

rhG-CSF is given starting 24 h after the cyclophosphamide, to augment neutrophil recovery and to mobilize HSCs into the circulation. It is administered daily for 11 days by subcutaneous injection according to the following weight-based nomogram.

Weight (kg)	Daily G-CSF dose (mcg)
<30	300
30–50	480
>50–60	600
>60–80	780
>80	960

The patient may self-administer the G-CSF once appropriate teaching has been provided.

4.3. Complications and Supportive Care

4.3.1. Monitoring

Cyclophosphamide can cause hemorrhagic cystitis. The urine is checked by dipstick for the presence of occult blood for every 8 h. If urine contains 2+ or 3+ occult blood, the MESNA infusion rate is increased to 70 ml/h.

High-dose cyclophosphamide induces an SIADH; serum electrolytes and serum creatinine are monitored daily for 3 days. Appropriate medical interventions are instituted should hyponatremia develop.

Blood counts and WBC differentials are routinely checked on 5th, 7th, 10th, and 11th day after Cyclophosphamide to determine the severity of cytopenias and kinetics of marrow recovery.

4.3.2. Nausea and Vomiting

Chemotherapy-induced nausea and vomiting, if it develops, may be treated with Prochlorperazine (10 mg po or iv) every 4 h, or Metoclopropamide (20–30 mg po or iv) every 4 h as required. Diphenhydramine hydrochloride (25 mg po or iv) may be given concurrently with the anti-emetics to prevent dystonic reactions. The anti-emetics may be alternated every 2 h if required for persistent nausea. Should the nausea persist, the frequency of prophylactic Ondansetron may be increased up to 8 mg po every 6 h as needed.

4.3.3. Cytopenias

Severe thrombocytopenia (platelets less than $20 \times 10^9/l$) develops infrequently but should be treated with prophylactic transfusion of donor leukocyte-reduced platelets. Severe anemia (hemoglobin less than 80 g/l) rarely develops but may be treated with prophylactic transfusion of 2 units of leukocyte-reduced packed red blood cells, if the patient is symptomatic from the anemia.

4.3.4. Febrile Neutropenia

Febrile neutropenia frequently develops following high-dose cyclophosphamide. Patients are taught to contact the BMT specialist on call (available 24 h a day, 7 days a week) if their temperature is greater than 38°C for 2 consecutive hours or at any time if it increases to 38.5°C. The febrile patient is admitted to hospital and a physical examination is performed looking for hemodynamic instability, and a source of infection, paying particular attention to the Hickman line exit site and urine. Blood counts and WBC differential, blood cultures, a urine culture, Hickman line exit site swab for culture, and a chest X-ray are obtained to determine the source of the fever. Piperacillin/Tazobactam (3.375 g iv q6h), Meropenem (500 mg iv q6h), or other appropriate broad-spectrum antibiotic are started immediately after the cultures are obtained. Infectious disease consultation should be obtained, particularly if there is evidence of septic shock or slow resolution of the fever.

4.3.5. Cytokine-Induced Bone Pain

Bone pain and myalgias resulting from cytokine administration may be treated with acetaminophen (650-mg q4h prn). Occasionally stronger narcotic analgesia is required.

4.4. Autologous Peripheral Blood Stem Cell Collection

Daily peripheral blood WBC and CD34 counts, starting 10 days after Cyclophosphamide, are used to predict the timing of stem cell collection (*18*). Leukopheresis is initiated on the day when the absolute circulating CD34 cell count rises to 10 CD34 cells/μl. The peripheral blood cell collection is typically performed on day 11 following chemotherapy.

Leukopheresis is carried out using a Cobe Spectra or other similar machine, by a nurse trained in the procedure. Venipuncture of the antecubital veins using 15 gauge needles are used for venous access and provide adequate blood flow in most patients. The Hickman catheter may be used for a return line. If venipuncture access is inadequate, a dialysis type catheter may need to be placed temporarily to allow adequate venous access. Blood is drawn from the patient and through the leukopheresis machine, collecting the WBC and returning the other blood components to the patients. Fifteen to twenty liters of blood is processed on each collection day. While a single leukopheresis is generally sufficient, this procedure is repeated daily until a minimum of 6×10 (*6*) CD34 cells/kg of recipient weight have been collected or until a maximum of three leukopheresis have been performed. Should more than one collection be required, the first is stored at 4°C overnight and pooled with the second collection for further processing.

4.5. Graft Processing

4.5.1. CD34 Selection of the PBSC Graft

Immune cell depletion of the graft is accomplished by positively selecting HSC away from other cells in the graft product using a Miltenyi CliniMACS Stem Cell Selection Device according to the manufacturer's specifications. The graft is incubated with a monoclonal antibody to CD34, an antigen expressed on HSC, linked to a nanoscale paramagnetic bead. The graft is passed through a column surrounded by magnet. The CD34-bound HSC are retained in the magnetic field while the remainder of the cells is washed away. The magnetic field is released and the purified HSC are collected.

4.5.2. Quality Testing of the CD34-Selected PBSC

Microbiological sterility, mononuclear cell counts, and CD34 cell counts of the graft product are measured as in **Subheading 3.2**. Quantitative flow cytometry, using appropriate fluorochrome-labeled monoclonal antibodies, is used to enumerate immune cell subpopulations (CD3/CD4+, CD3/CD8+, CD19+, CD56+, and CD14+) in the graft product before and after CD34 cell selection. Typically, about 50% of the stem cells are recovered from the CD34 selection process and upward of 97% of the cells in the selected graft express CD34.

If the CD34-selected peripheral blood stem cell (PBSC) has less than 2×10^6 CD34 cells per kg of recipient weight, the patient will be ineligible for further treatment on this protocol. The stem cell graft will be deemed appropriate for use in this study if, after CD34 cell selection, it contains less than 5×10^4 T lymphocytes per kg of recipient weight (*5*).

4.5.3. Cryopreservation of the CD34-Selected PBSC

The CD34-selected PBSC are divided into two equal aliquots; mixed with an equal volume of cryopreservation solution (*3*) and cryopreserved in two separate bags in a similar fashion to the bone marrow graft (**Subheading 3.3**) in the vapor phase of liquid nitrogen until transplantation.

5. High-Dose Immune Ablative Therapy and HSCT

Busulphan, cyclophosphamide, and antithymocyte globulin are used to destroy the autoreactive immune system in the MS patient. HSC in the CD34-selected PBSC is used to replace the bone marrow stem cells killed as a consequence of the chemotherapy. The high-dose therapy used for immune ablation causes a number of predictable side-effects which are treated prophylactically. Recently published overviews of stem cell transplantation list the common side-effects of this procedure (*19, 20*). Vigilant medical care by a multidisciplinary team expert in HSCT is important for early detection of a myriad of complications and prompts institution of the appropriate therapy. This phase of the protocol is carried out with the patient admitted to the transplant unit.

5.1. High-Dose Busulphan Administration

High-dose busulphan penetrates into the central nervous system (CNS); while this is beneficial for immune ablation of CNS immunity, this drug has been associated with seizures. Antiseizure prophylaxis is used during its administration.

5.1.1. Antiseizure Prophylaxis

Patients are started on Phenytoin (300 mg po) 24 h prior to receiving Busulphan. The first three doses of Phenytoin are given every 8 h, to rapidly achieve therapeutic serum levels. It is then continued once daily for the duration of the busulphan treatment, being discontinued following the last dose of Busulphan. Patients with a hypersensitivity to Phenytoin receive an alternative antiseizure medication.

5.1.2. Antiemetic Prophylaxis

Busulphan is moderately emetogenic. Antiemetic prophylaxis, with Ondansetron (8 mg po or iv) or an equivalent HT_3 receptor blocker, is started prior to the first dose of Busulphan and continue for 24 h following the last dose of busulphan.

5.1.3. Busulphan

Busulphan (0.6 mg/kg in 100 ml NS) is given by intravenous infusion over 2 h every 6 h for 16 doses. The Busulphan dose is based upon the lesser of the patient's ideal or actual body weight.

5.1.4. Busulphan Pharmacokinetics

Busulphan pharmacokinetics is able to predict the risk of veno-occlusive disease associated with this drug (*21*) and is used to adjust the dosing to reduce the risk of this complication. First dose pharmacokinetics parameters are calculated from plasma concentrations in samples drawn hourly after the first dose. The blood sample is drawn into a 3.0 ml heparinized blood tube and immediately put on ice. The plasma is separated by centrifugation of the blood at $2,000 \times g$ for 20 min at 4°C and decanted to a fresh tube. Busulphan is measured by HPLC against a known set of standard samples. The Busulphan area under the curve (AUC – concentration vs. time) is calculated by the trapezoidal method. If the Busulphan first dose AUC is greater than $1,200$ μmol × min/l, Busulphan dose 5–16 will be proportionately reduced to give an estimated AUC = $1,200$ μmol × min/l.

5.2. High-Dose Cyclophosphamide Administration

Busulphan induces hepatic enzymes that increase the conversion of cyclophosphamide to its active form which may result in increased toxicity with the latter drug. Cyclophosphamide will not begin for a minimum of 24 h following the last dose of Busulphan.

5.2.1. Supportive Care

Hyperhydration (**Subheading 4.1.1**), Foley catheter bladder drainage (**Subheading 4.1.1**), MESNA (**Subheading 4.1.2**), and antiemetics (**Subheadings 4.1.3** and **4.3.2**) are started as previously outlined and continued until 24 h following the administration of the final dose of cyclophosphamide.

5.2.2. Cyclophosphamide

Cyclophosphamide (50 mg/kg in 1,000 ml of NS) is given as a 1 h intravenous daily dose for 4 consecutive days. The Cyclophosphamide dose is based upon the lesser of the patient's actual or ideal body weight.

5.3. Rabbit Antithymocyte Globulin

Rabbit Antithymocyte Globulin (rATG) frequently causes fever and anaphylactoid reactions. To prevent these infusional side-effects, the patient is premedicated with Diphenhydramine (25 mg po or iv) 30 min and Methylprednisolone (1 mg/kg iv) 30 min prior to the infusion of each dose of rATG. The Methylprednisolone is repeated 12 h after each dose of rATG. Vital signs are monitored every 15 min during the infusion of rATG and for 30 min following the infusion. Febrile reactions are treated with Acetaminophen (650 mg po q4h) as required. Other reactions are treated with appropriate supportive measures.

rATG (1.25 mg/kg/in 250-ml NS) is given through an inline 0.22-µm filter into a central venous catheter over 4 h as a single daily intravenous infusion for 4 consecutive days. Patients sensitized to rATG or who experience a severe reaction while receiving rATG, will have equine ATG (at a dose of 5.0 mg/kg) substituted for the remaining doses of rATG.

5.4. Stem Cell Reinfusion

The autologous CD34-selected PBSC graft is administered 48 h after the last dose of Cyclophosphamide. Prochlorperazine (10 mg po or iv) is given to the patient about 15 min prior to infusing the graft to prevent nausea associated with the graft reinfusion. One of the two cryobags containing the CD34-selected PBSC graft is removed from liquid nitrogen storage and thawed rapidly at the bedside in a 37°C water bath. The thawed CD34-selected PBSC stem cell graft is infused into the patient through the central venous line over 20 min. The second cryobag is then thawed and infused.

5.5. Complications and Supportive Care

5.5.1. Monitoring

Because of the complex nature of HSCT and the critical nature of the side-effects of the immune ablative therapy, patients undergoing this procedure require close observation. Vital signs are performed three times daily, but more frequently as required. The patient's weight is recorded daily. Routine daily laboratory test include complete blood counts and differential, serum electrolytes and creatinine. A coagulation profile is monitored twice weekly. Serum albumin, calcium, phosphate, liver function tests, total protein, magnesium are routinely measured twice weekly. A type and screen is drawn twice weekly.

5.5.2. Nausea and Vomiting

Chemotherapy-induced nausea and vomiting are treated as previously outlined (**Subheading 4.3.2**).

5.5.3. Prevention of Urate Nephropathy

Allopurinol (300 mg po) is administered daily starting prior to the first dose of chemotherapy and continuing to the day of HSCT.

If urate nephropathy develops, patients are managed with urine alkalinization and hyper-hydration with forced diuresis.

5.5.4. Veno-Occlusive Disease Prophylaxis

Veno-Occlusive Disease (VOD) is a syndrome resulting from the hepatic toxicity of the high-dose Busulphan and Cyclophosphamide. It is manifest as weight gain, tender hepatomegaly, and hyperbilirubinemia. It may progress to include capillary leak syndrome resulting in anasarca, noncardiogenic pulmonary edema, and respiratory failure and renal failure. Enoxaparin (20 mg sc/day) and Ursodiol (250 mg po tid) are given to reduce the risk of hepatic injury. Enoxaparin is administered from the day prior to the start of the preparative chemotherapy until the patient is discharged from hospital or day + 30 post-HSCT, whichever occurs first. This drug is discontinued for any major bleeding episode. Ursodiol is given starting 24 h prior to the first dose of busulphan and continued for 90 days following HSCT.

5.5.5. Mucositis

Oral hygiene is maintained by frequent rinsing of the mouth using saline mouthwashes. Mucositis may cause severe pain which is controlled with appropriate doses of intermittent oral or intravenous narcotics or the use of a transdermal Fentanyl.

5.5.6. Nutritional Support

Frequently, chemotherapy-induced effects on the digestive tract limit the nutritional intake of the patient. Patients who are unable to swallow receive enteral nutrition via a soft nasogastric feeding tube until a sustained oral intake of at least 1,500 kcal/day is achieved. Total parental nutrition (TPN) is used for patients unable to tolerate nasogastric feeding.

5.5.7. Cytopenias

Neutropenia and thrombocytopenia are severe and typically last for about 2 weeks following HSCT. Transfusion support is given for cytopenias as previously outlined (**Subheading 4.3.3**). Oral contraception is used to suppress menses during the period of severe thrombocytopenia in female patients.

rhG-CSF

rhG-CSF (5 μg/kg sc daily) is given beginning on the day following HSCT, in order to minimize the duration of neutropenia. It is continued until neutrophil engraftment, the second consecutive day that the neutrophil count is greater than $0.5 \times 10^9/l$.

5.5.8. Infection Prophylaxis

Patients are globally immune deficient following chemotherapy. Immune reconstitution occurs over 18 months following HSCT. Prophylaxis is used to minimize the risk of opportunistic infections that may commonly occur during this period of risk.

Mucocutaneous Candidiasis

Patients receive Fluconazole (400 mg po daily) to prevent mucocutaneous candidiasis. It is started 4 days prior to HSCT until the recovery of the neutrophil count following HSCT. At this point, the Fluconazole dose is reduced to 100 mg po daily and continued for 90 days post-HSCT. Because of regional differences in the etiologic agents responsible for opportunistic infections, fungal prophylaxis should be tailored upon the microbiological flora of the transplant center.

Pneumocystis Carinii Pneumonia Prophylaxis

Pneumocystis Carinii Pneumonia (PCP) prophylaxis is achieved with Trimethoprim/Sulfamethoxasole (160 mg/800 mg) given orally twice daily on Monday and Thursday every week starting when the neutrophil count rises above $0.5 \times 10^9/l$ and continuing for 12 months following HSCT. In the event of a Trimethoprim or sulfa allergy, inhalational Pentamidine 300 mg monthly is used as an alternate PCP prophylaxis.

Intravenous Gammaglobulin

Passive immunity is provided by prophylactic intravenous gammaglobulin (IVGG) infusions which start 7 days following HSCT and continue throughout the first post-HSCT year. Initially, IVGG (500 mg/kg iv) is given weekly. Serum IgG levels are measured just prior to the infusion of each dose of the IVGG and used to adjust the frequency of administration of subsequent doses of IVGG.

Herpes Simplex Virus, Cytomegalovirus, and Varicella-Zoster Virus Prophylaxis

Herpes simplex virus (HSV) stomatitis frequently occurs during the immediate post-HSCT period, while Cytomegalovirus (CMV) reactivation causing pneumonia may occur in the first 4 months post-HSCT. Shingles may develop due to Varicella-Zoster Virus (VZV) reactivation, generally between 3 and 6 months post-HSCT. All patients are given Valacyclovir (2 g po three times daily) from 4 days prior to HSCT until day + 100 post-HSCT to minimize the risk of HSV and CMV reactivation (*22, 23*). Valacyclovir is continued at a lower dose (0.5 g twice a day) for 6 months post-HSCT to prevent shingles. Patients may be given acyclovir (500 mg/m² i.v. tid) if they are unable to take oral medications.

PCR detection of CMV is performed weekly for the first 3 months post-HSCT to detect reactivation of the virus. Viremia is treated using intravenous Gancyclovir.

5.5.9. Febrile Neutropenia

Febrile neutropenia, bacteremia, and sepsis are frequent events during HSCT and are treated as previously outlined (**Subheading 4.3.4**).

6. Postengraftment Care

The patient is discharged from the hospital following recovery of cytopenias, when hydration and nutritional intake are adequate and following resolution of acute HSCT complications.

Frequent medical checkups are required, typically twice weekly for the first several months following discharge, to monitor the blood count recovery, hydration and nutritional status of the patient, and for the development of opportunistic infections.

6.1. Central Venous Catheter Removal

Catheters are removed under local anesthetic at the exit site. The cuff is separated from the subcutaneous tissue by blunt dissection through the exit site. Once the cuff is free, the catheter is removed. An occlusive dressing is placed over the exit site for 3 or 4 days.

6.2. Posttransplant immunization

Immune memory, including memory to childhood vaccines, is abrogated by HSCT. Immune reconstitution proceeds over the first year following transplantation. Patients will receive a primary immunization series against Diphtheria, Tetanus, Polio (using inactivated polio vaccine), Hemophilus influenza, Pneumoccocus, and Hepatitis B. Patients are advised to receive the Influenza Vaccine yearly in the fall. Immunization with live virus vaccines may be performed following advice from an infectious disease specialist but are deferred for at least 2 years post-HSCT. Patients who have a history of hypersensitivity to a vaccine or a component of the vaccine will not be immunized with that vaccine.

The Sabin oral polio vaccine (OPV) should not be given to family members within the first year after transplantation in order to protect the HSCT recipient. It would be advisable for family members to receive inactivated (Salk) polio vaccine.

7. Notes

1. BEAM is an acronym for combination chemotherapy regimen composed of Carmustine (BCNU), Etoposide, Cytarabine (Ara-C), and Melphalan

2. ACD-A refers to Anticoagulant Citrate Dextrose Solution A. It is made by several manufacturers.

3. The cryoprotectant solution is composed of 25 ml of Dimethyl Sulfoxide, 108 ml of Pentaspan, 80 ml of 25% (v/v) of human serum albumin, and 37 ml of Plasma-Lyte A (Baxter) for a total volume of 250 ml. The solution is made and cooled to 2–8°C just prior to each graft cryopreservation using aseptic techniques.

4. Peripherally inserted central catheters (PICC) provide an alternative method of vascular access in bone marrow transplant recipients (*24*)

5. If the CD34-selected PBSC contains more than 5×10^4 T lymphocytes per kg of recipient weight, T cell depletion of the CD34-selected peripheral blood stem cell graft may be performed using an approved available method in accordance with stem cell processing laboratory's standard operating procedure in an attempt to reduce the number of residual T cells below the threshold of 5×10^4 T lymphocytes per kg of recipient weight.

References

1. Burt RK, Burns W, Ruvolo P, et al. (1995). Syngeneic bone marrow transplantation eliminates V beta 8.2 T lymphocytes from the spinal cord of Lewis rats with experimental allergic encephalomyelitis. *J Neurosci Res*, **41**:526–531.

2. van Gelder M, Kinwel-Bohre EP, van Bekkum DW. (1993). Treatment of experimental allergic encephalomyelitis in rats with total body irradiation and syngeneic BMT. *Bone Marrow Transplant*, **11**:233–241.

3. Fassas A, Anagnostopoulos A, Kazis A, et al. (1997). Peripheral blood stem cell transplantation in the treatment of progressive multiple sclerosis: first results of a pilot study. *Bone Marrow Transplant*, **20**:631–638.

4. Atkins H and Freedman M. (2005). Immunoablative therapy as a treatment aggressive multiple sclerosis. *Neurol Clin*; **23**:273–300, ix.

5. Freedman MS and Atkins HL. (2004). Suppressing immunity in advancing MS: too much too late, or too late for much? *Neurology*; **62**:168–169.

6. Gratwohl A, Passweg J, Bocelli-Tyndall C, et al. (2005). Autologous hematopoietic stem cell transplantation for autoimmune diseases. *Bone Marrow Transplant*; **35**:869–879.

7. Personal Communication.

8. Nash RA, Bowen JD, McSweeney PA, Pavletic SZ, Maravilla KR, Park MS, Storek J, Sullivan KM, Al-Omaishi J, Corboy JR, DiPersio J, Georges GE, Gooley TA, Holmberg LA, LeMaistre CF, Ryan K, Openshaw H, Sunderhaus J, Storb R, Zunt J, Kraft GH. (2003). High-dose immunosuppressive therapy and autologous peripheral blood stem cell transplantation for severe multiple sclerosis. *Blood*; **102**(7):2364–2372.

9. Burt RK, Cohen BA, Russell E, Spero K, Joshi A, Oyama Y, Karpus WJ, Luo K, Jovanovic B, Traynor A, Karlin K, Stefoski D, Burns WH. (2003). Hematopoietic stem cell transplantation for progressive multiple sclerosis: failure of a total body irradiation-based conditioning regimen to prevent disease progression in patients with high disability scores. *Blood*; **102**(7):2373–2378.

10. Aldouri MA, Ruggier R, Epstein O, Prentice HG. (1990). Adoptive transfer of hyperthyroidism and autoimmune thyroiditis following allogeneic bone marrow transplantation for chronic myeloid leukaemia. *Br J Haematol*; **74**(1):118–119.

11. Wyatt DT, Lum LG, Casper J, Hunter J, Camitta B. (1990). Autoimmune thyroiditis after bone marrow transplantation. *Bone Marrow Transplant*; **5**(5):357–361.

12. Fassas A, Anagnostopoulos A, Kazis A, Kapinas K, Sakellari I, Kimiskidis V, Smias C, Eleftheriadis N, Tsimourtou V. (2000). Autologous stem cell transplantation in progressive multiple sclerosis – an interim analysis of efficacy. *J Clin Immunol*; **20**(1):24–30.

13. Weinshenker BG. (1994). Natural history of multiple sclerosis. *Ann Neurol*; **36**:6–11.

14. Adami GF, Bacigalupo A, Bonalumi U, Van Lindt MT, Griffanti-Bartoli F. (1981). Use of Hickman right atrial catheter for vascular access in marrow transplant recipients. *Arch Surg*; **116**(8):1099.

15. Hickman RO, Buckner CD, Clift RA, Sanders JE, Stewart P, Thomas ED. (1979). A modified right atrial catheter for access to the venous system in marrow transplant recipients. *Surg Gynecol Obstet*; **148**(6):871–875.

16. Sutherland DR, Anderson L, Keeney M, Nayar R, Chin-Yee I. (1996). The ISHAGE guidelines for CD34+ cell determination by flow cytometry. *International Society of Hematotherapy and Graft Engineering. J Hematother*; **5**(3):213–226.

17. http://www.halls.md/body-surface-area/bsa.htm.

18. Mitterer M, Hirber J, Gentilini I, Prinoth O, Fabris P, Emmerich B, Coser P, Straka C. (1996). Target value tailored (TVT) apheresis approach for blood progenitor cell collection after high-dose chemotherapy and rh-G-CSF. *Bone Marrow Transplant*; **18**(3):611–617.

19. Copelan EA. (2006). Hematopoietic stem-cell transplantation. *N Engl J Med27*; **354**(17):1813–1826.

20. Leger CS, Nevill TJ. (2004). Hematopoietic stem cell transplantation: a primer for the primary care physician. *Can Med Assoc J*; **170**(10):1569–1577.

21. Dix SP, Wingard JR, Mullins RE, Jerkunica I, Davidson TG, Gilmore CE, et al. (1996). Association of busulfan area under the curve with veno-occlusive disease following BMT. *Bone Marrow Transplant*; 17:225–230.

22. Lowance D, Neumayer HH, Legendre CM, Squifflet JP, Kovarik J, Brennan PJ, et al. (1999). Valacyclovir for the prevention of cytomegalovirus disease after renal transplantation. *International Valacyclovir Cytomegalovirus Prophylaxis Transplantation Study Group. N Engl J Med*; **340**:1462–1470.

23. Feinberg JE, Hurwitz S, Cooper D, Sattler FR, MacGregor RR, Powderly W, et al. (1998). A randomized, double-blind trial of valaciclovir prophylaxis for cytomegalovirus disease in patients with advanced human immunodeficiency virus infection. *AIDS Clinical Trials Group Protocol 204/Glaxo Wellcome 123–014 International CMV Prophylaxis Study Group. J Infect Dis*; 177:48–56.

24. McDiarmid S, Hamelin L, Huebsch LB. (2006). Leading change: Retrospective evaluation of a nurse-led initiative in vascular access options for autologous stem cell transplant recipients ranging from Hickman catheters to peripherally inserted central catheters. *J Infus Nurs*; **29**(2):81–88.

INDEX

A

Adipogenic induction 113
Adrenoleukodystrophy (ALD) 176
Alzheimer's disease.. 147
Amyotrophic lateral sclerosis (ALS)............................... 24

B

Basal culture media.. 87
Blood–brain barrier (BBB) 158, 161
Bone marrow-derived cells 18–19
Bone marrow graft
    cryopreservation.. 234
    harvest 233–234
    quality testing 234
Bordetella pertussis...158, 165
5-Bromo-2 -deoxyuridine (BrdU) labelling
    chromogen-based immunohistochemistry....... 223–226
    fluorescence-based immunohistochemistry 226–227
    materials 219–220

C

Caenorhabditis elegans...143, 144
CAG. *See* Cysteine, adenine and guanine
Callithrix jacchus..163–164
CAPIT. *See* Core assessment protocol for intracerebral
    transplantation
CCA. *See* Common carotid artery
Cell identification
    cellular aggregation.. 70
    immunostaining... 70–71
    oligodendrocyte progenitors
        markers 71
        morphology 70
    stem cell growth
        markers 70
        morphology 69–70
    yellow spheres/neural progenitors
        markers 71
        morphology 70
Cellular imaging
    extracellular cell labelling
        CD34+ cells... 203
        CliniMACS CD34+ selection technology
            (Miltenyi) 204
        immunomagnetic selection 202–203

intracellular cell labelling
        D-mannose cell coating..................................... 202
        PC-SPIO nanoparticles............................. 200–201
        SPIO stabilization 199
Central nervous system (CNS)................21, 157, 158, 163
Chemistry and manufacturing controls (CMC)............... 6
Chondrogenic induction... 114
Chromogen-based BrdU immunohistochemistry
    blocking and antibody steps............................ 224–225
    counterstaining 225–226
    free-floating sections, slides 225
    wax and frozen section pre-treatment...................... 224
CMV. *See* Cytomegalovirus
Common carotid artery (CCA)............................. 190, 191
Core assessment protocol for intracerebral
    transplantation (CAPIT)................................ 44
Cyclophosphamide administration
    anti-emetic prophylaxis.. 236
    dosage... 236, 241
    hyperhydration.. 236
    risk management 235–236
    sodium 2-mercaptoethane sulfonate (Mesna).......... 236
    supportive care.. 240
Cysteine, adenine and guanine (CAG)...........137, 142, 143
Cytomegalovirus (CMV) 149

D

Dania rerio..143
Delayed type hypersensitivity (DTH) 162
Density analyzer program... 131
Diffusion weighted images (DWI)........................ 192, 193
Drosophila melanogaster..143, 144
Dulbecco's Modified Eagle Medium (DMEM).... 208, 210

E

ECA. *See* External carotid artery
Embryonic stem cell (ESC)
    autologous *vs.* allogeneic cell therapy...................... 4, 5
    and bone marrow cell fusion 18–19
    treatment 20, 21
Epidermal growth factor (EGF)..................................... 80
Expanded Disability Status Score (EDSS)............ 232–233
Experimental autoimmune encephalomyelitis (EAE)
    clinical evaluation 166, 168
    inbred species
        BBB migration 161

Experimental autoimmune encephalomyelitis (EAE)
(*Continued*)
encephalitogenic and nonencephalitogenic
T cells 160
MHC molecules 161–162
T cell receptor (TCR) 160–161
Th1 and Th2 cytokines............................ 162–163
induction
materials 165
methods 166, 167
outbred species.. 163–164
External carotid artery (ECA) 189, 190

F

Fast low angle gradient echo (FLASH) sequence 212
Fibroblast growth factor (FGF) 80
Flow cytometric characterization, neural precursor cell
central nervous system (CNS) cell 78
continued and robust neurogenesis 77
materials
antibodies 82
commercial media preparation 79–80
dissection instrument 79
in-house media preparation 80–82
tissue culture equipment 79
methods
flow cytometric enrichment 86–87
setup 83
SVZ tissue dissection 83–85
FLOWJO analysis software 126
Fluorescence-based BrdU immunohistochemistry
blocking and antibody steps 226–227
wax and frozen section pre-treatment 226
Fluorescent activated cell sorting 86

G

Galactocerebrosidase gene 176
Glial cells 52, 57
Gliogenesis 217
Globoid cell leukodystrophy. *See* Krabbe's disease
Good manufacturing practicable (GMP) process 14
Green flourescent protein (GFP) 149
Growth media preparation
commercial
complete NSC media 80
culture reagents 79
growth factor stock solutions 80
hormone supplemented growth media 80
in-house
culture reagents 80–81
stock solutions 81–82

H

Hank's balanced salt solution (HBSS) 122, 131
Hanks eagle medium (HEM) 82

HD. *See* Huntington's disease
Hematopoietic stem cell (HSC)
neurological disease 17–18
scale-up and manufacturability 6
Hematopoietic stem cell transplantation
(HSCT) 232, 233
Heparin 80
Human embryonic stem cells (hESC)
differentiation 9
materials
recipes 61–63
stock media/solutions 61
stock solutions and medium storage 53
supplements 61, 62
supplies 52, 61
methods
cellular aggregation 66–67
definitive neuroepithelia isolation 56–57
ES aggregates/"embryoid bodies" 54–55
oligodendrocyte progenitors 68–69
stem cell growth 65–66
vs. primitive and definitive
neuroepithelia 55–56
yellow sphere formation 67–68
tumorigenic stroma cells 52
Human immunodeficiency virus (HIV) 149
Human mesenchymal stem cells (hMSC) culture
bone marrow extraction 107–108
coverslip growth 110
immunostaining
coverslip-grown cultures 115–116
materials 107
NSC-like neurosphere suspension 116
lab equipment 104–105
long-term storage
cryopreservation 110
materials 106
thawing 111
neurogenic capacity
intact fixation 115
materials 106–107
neural differentiation 115
NSC-like neurosphere generation 114–115
origin confirmation
adipogenic induction 113
chondrogenic induction 114
flow cytometry 111–113
materials 106
osteogenic induction 113–114
passage 109–110
primary antibodies 112–114, 116
Human Tissue Act (HTA) 42, 43
Huntington's disease (HD)
excitotoxic models
bregma 140–141
sagittal section photomicrograph 139, 140

stereotactic apparatus 139, 141
striatum 139–140
genetic mouse models
knock-in models 143
transgenic mouse models 142–143
non-mouse models 143–144
stem cells 23
transplantation
donor tissue 148
host site 150
labelling of cells 148–150
procedure 150

I

ICA. *See* Internal carotid artery
Immune ablation, multiple sclerosis
HSC mobilization and collection
complications and care 237–238
cyclophosphamide administration 235–236
graft processing 238–239
peripheral blood cell collection 238
rhG-CSF 237
and HSCT
busulphan administration 239–240
complications and care 241–243
cyclophosphamide administration 240–241
rabbit antithymocyte globulin (rATG) 241
patient selection 232–233
postengraftment care 243–244
preparative phase
bone marrow graft cryopreservation
and testing 234
bone marrow harvest 233–234
central venous catheter care 235
vascular access and postoperative analgesia 234
Immunocytochemistry
materials 120–121
methods 124–125
Immunostaining
mesenchymal stem cell (MSC) 107, 115–116
oligodendrocyte precursor cell (OPC) 71–72
progenitor and neural stem cells 100–101
Independent review board (IRB) 43–44
Injured brain extracts, UCB migration
materials 121
methods
cell preparation 127
migration assay 127–129
tissue extract preparation 127
Internal carotid artery (ICA) 189–191
Internal ribosomal entry site (IRES) 149
I.V. transplantation
infusion
femoral vein 129
jugular vein 129–130

penile vein 130
tail vein 129
UCB cell preparation 121, 129

K

Ki67 anitbody 220, 227
Krabbe's disease 176

L

LacZ gene encoding 149–150
Leukaemia inhibitory factor (LIF) 34
Leukodystrophy
ALD and MLD 176
culture, mouse neonates
materials 177
methods 178–179
historical evaluation 182
oligosphere dissociation
materials 178
methods 179–180
Pelizaeus–Merzbacher disease (PMD) 175
Schwann cells 177
transplantation
materials 178
methods 180–182

M

Magnetic resonance imaging (MRI), cell tracking
cell culturing
materials 208–209
methods 210–211
cell labelling
materials 209
methods 211
cellular imaging
contrast agents 198, 199
extracellular cell labelling 202–204
intracellular cell labelling 199–202
implanted cells
brain tumors 207–208
injured brain 204, 205
iron fluorophore particles 208
neurodegenerative disorders 207
spinal cord injury (SCI) 204–207
Prussian blue stain
materials 209
methods 212
in vitro
materials 209
methods 211
in vivo
materials 209
methods 211–212
MBP. *See* Myelin basic protein

Medicines and Healthcare products
       Regulatory Agency (MHRA) 41
Mesenchymal stem cell (MSC)
   human culture
      materials 105
      methods 107–110
   immunostaining
      materials 107
      methods 115–116
   long-term storage
      materials 106
      methods 110–111
   neurogenic capacity
      materials 106–107
      methods 114–115
   origin confirmation
      materials 106
      methods 111–114
Metachromatic leukodystrophy (MLD) 176
Middle cerebral artery occlusion (MCAO) model 126
   induction
      materials 188
      methods 190, 191
   infarct volume, TTC staining and quantitative
      analysis
      materials 189
      methods 191–192
   intravenous delivery
      materials 189
      methods 191
   magnetic resonance imaging (MRI)
      materials 189
      methods 192–193
MSC. *See* Mesenchymal stem cell
Multicolor flow cytometry
   materials 121
   methods 125–126
Multi-label fluorescence staining
   blocking and antibody steps 227–228
   frozen section pre-treatment 227
Multiple sclerosis (MS)
   EAE induction
      inbred species 159–163
      materials 165
      methods 166, 167
      outbred species 163–164
   HSC mobilization and collection
      complications and care 237–238
      cyclophosphamide administration 235–236
      graft processing 238–239
      peripheral blood cell collection 238
      rhG-CSF 237
   and HSCT
      busulphan administration 239–240

      complications and care 241–243
      cyclophosphamide administration 240–241
      rabbit antithymocyte globulin (rATG) 241
   MOG and PLP 158, 159
   myelin basic protein (MBP) 158
   patient selection 233–234
   postengraftment care 243–244
   preparative phase
      bone marrow graft cryopreservation
         and testing 234
      bone marrow harvest 233–234
      central venous catheter care 235
      vascular access and postoperative
         analgesia 234
   SCH preparation
      materials 165
      methods 165–166
   T lymphocytes 158
Mycobacterium tuberculosis 158, 165, 166
Myelin basic protein (MBP)
   hydrophilic myelin protein 158
   inbred species 160–162
   outbred species 164
Myelin oligodendrocyte glycoprotein (MOG)
   humoral and cellular immune response 158
   immunization 159, 164

N

Neural induction medium
   definitive neuroepithelia isolation 56
   hESC *vs.* primitive and definitive
      neuroepithelia 55–56
   materials, hESC 53
Neural precursor cell culture, flow cytometric
      characterization
   central nervous system (CNS) cell 78
   continued and robust neurogenesis 77
   materials
      antibodies 82
      commercial media preparation 79–80
      dissection instrument 79
      in-house media preparation 80–82
      tissue culture equipment 79
   methods
      flow cytometric enrichment 86–87
      setup 83
      SVZ tissue dissection 83–85
Neural progenitor cells (NPC)
   brain tisssue preparation
      fixation 222
      frozen brain processing and sectioning 223
      materials 218–219
      paraffin wax processing
         and sectioning 222–223

5-bromo-2 -deoxyuridine (BrdU) labelling
  chromogen-based
    immunohistochemistry 223–226
  fluorescence-based
    immunohistochemistry 226–227
  materials 219–220
common antibodies 220–221
detection 228
materials
  recipes 61
  supplies and solutions 60
methods
  growth and feeding spheres 64
  isolation and plating 64
  neurosphere plating 65
  transplantation 65
multi-label fluorescence staining 227–228
sub-granular zone (SGZ) 218
sub-ventricular zone (SVG) 217, 218
Neural stem cells (NSC)
  autologous *vs.* allogeneic cell therapy 4–5
  cell number determination 10–11
  clinical end point and patient recruitment
    criteria 13–14
  differentiation methods 3, 4
  good manufacturing practicable (GMP) process 14
  mode of action assessment 11–12
  neurosphere
    generation 114–115
    intact fixation 115
    suspension 116
  scale up and manufacturability
    cell manufacturing protocols 10
    hematopoietic stem cells (HSC) 6
    human embryonic stem cells (hESC)
      differentiation 9
    human *vs.* rodent stem cells 7–8
    stem cell-based transplant therapy 6
Neurodegenerative diseases
  CAPIT-HD and CAPIT-PD protocols 44
  clinical trials
    bone marrow-derived stromal cells 39
    cell preparation 40–41
    cell replacement approach 37
    human foetal mesencephalic tissue 38
    PD and HD 38, 39
    xenotransplantation 39
  European Union (EU) directives
    good clinical practice (GCP) 41
    human tissue and cells 42–43
  Helsinki and GCP declaration 43
  independent review board (IRB) 43–44
  materials
    6-hydroxydopamine rodent model 138

striatal excitotoxic lesion 138
transplantation 138–139
methods
  excitotoxic models 139–142
  genetic mouse models 142–143
  6-hydroxydopamine rodent model 144–146
  non-mouse models 143–144
  α-synuclein transgenic mice 146–147
  transplantation 148–150
neurological disease 22–23
quality control 44
stem and precursor cells
  dopaminergic neurons differentiation 35
  embryonic stem (ES) cell 34
  induced pluripotent stem (iPS) cells 37
  neomycin resistance gene 36
  tyrosine hydroxylase (TH) positive
    neurons 35, 36
Neuroepithelial differentiation method 52
Neurogenesis 217
Neurological disease
  accessibility 22
  bone marrow-derived cells 18–19
  cell fusion 19
  haemopoietic stem cell (HSC) 17–18
  neurodegenerative disease
    amyotrophic lateral sclerosis (ALS) 24
    Huntington's disease (HD) 23
    multiple sclerosis 23–24
    Parkinson's disease (PD) 22–23
    stroke 24–25
  non-canonical stem cell repair 20
  polyploid cells 19
  stem cell treatment
    accessibility 22
    delivery 21–22
    safety 20–21
Neurosphere assay (NSA)
  EGF-response 92, 93
  materials
    basal and complete media 95
    general and dissection equipment 94
    growth and passaging ingredients 95
    NS media 94–95
    tissue culture equipment 94
  methods
    cell differentiation 98–99
    culture establishment 96–97
    culture passaging 97–98
    immunostaining 100–101
Neurosphere culture
  dissociated cell differentiation 99
  establishment
    dissociation protocol 97

Neurosphere culture (*Continued*)
    periventricular region dissection 96–97
    setup 96
    passaging 97–98
    whole cell differentiation 98–99
Nurr1 expression.. 35, 36

O

Oligodendrocyte progenitor cells (OPC)
    cell identification
        markers 70–72
        morphology 69–70
    culture, mouse neonates
        materials 177
        methods 178–179
    human embryonic stem cells
        materials 61–63
        methods 65–69
    markers and dilution... 63
    neonatal brain 180
    neural progenitor cells
        materials 60–61
        methods 64–65
    oligosphere dissociation
        materials 178
        methods 179–180
    spinal cord white matter 180–182
    tissue culture 72–73
    young/adult mice brain 180
Organotypic hippocampal culture
    materials 122
    methods 131
Osteogenic induction.................................... 113–114

P

Parkinson's disease (PD)
    6-hydroxydopamine rodent model
        bilateral lesions 145
        dopaminergic cells 144–145
        striatal dopamine depletion............................... 146
        unilateral lesions 145–146
    stem cells 22–23
    α-synuclein transgenic mice 146–147
    transplantation
        donor tissue.. 148
        host site.. 150
        labelling of cells 148–150
        procedure 150
Parkinson's therapy 11
Pasteur pipette technique 57
PCNA. *See* Proliferating cell nuclei antibody
PC-SPIO. *See* Polycation-bound superparamagnetic
    iron oxide
Pelizaeus–Merzbacher disease (PMD) 175, 176

Photochemical lesion.. 204, 205
Polycation-bound superparamagnetic
    iron oxide (PC-SPIO) 200–202
Polyploid cells.. 19
Progenitor and neural stem cells
    EGF-response 92, 93
    materials
        basal and complete media 95
        general and dissection equipment........................ 94
        growth and passaging ingredients........................ 95
        NS media... 94–95
        tissue culture equipment 94
    methods
        cell differentiation................................... 98–99
        culture establishment 96–97
        culture passaging.................................... 97–98
        immunostaining............................... 100–101
Proliferating cell nuclei antibody (PCNA) 220, 227
Proteolipid protein (PLP)..................................... 158, 160

R

Rabbit antithymocyte globulin (rATG)........................ 241
Recombinant human granulocyte-colony
    stimulating factor (rhG-CSF).............. 235, 237

S

Schwann cells 177
Serum-free culture system. *See* Neurosphere assay
SGZ. *See* Sub-granular zone
Smooth muscle actin (SMA).. 71
SOX10 protein 71
Spastic paraplegia type II (SPG II) 175
Spinal cord homogenate (SCH)............................159, 160,
    165, 166
    immunization 159–160
    preparation
        materials 165
        methods 165–166
Spinal cord injury (SCI), MRI tracking
    balloon-induced compression lesion 204–205
    olfactory ensheathing glia (OEG)
        migration 206–207
        Prussian blue staining 206
Stem and precursor cells
    dopaminergic neurons differentiation........................ 35
    embryonic stem (ES) cell.. 34
    induced pluripotent stem (iPS) cells......................... 37
    neomycin resistance gene.. 36
    Nurr1 expression... 35, 36
    tyrosine hydroxylase (TH) positive
        neurons 35, 36
Stem cell-based transplant therapy............................. 6
Stereomicroscope.. 177, 178
Sub-granular zone (SGZ) 218

Sub-ventricular zone (SVZ) region
    adult brain 217, 218
    dot plot scatter analysis... 87
    hippocampal formation 77
    neural precursor cell enrichment.............................. 86
    tissue dissection 83–85
Superparamagnetic iron oxide (SPIO)
        nanoparticles
    cell labeling...199, 209, 211
    MR contrast agents 198
    resovist agent 208

T

Tissue culture 72–73
2,3,5-Triphenyltetrazolium chloride (TTC).......... 189, 192
Trituration88
Trypan blue dye exclusion method 125, 127, 129
Trypsin-EDTA...83, 85, 88
Tumorigenic stroma cells..52
T_2-weighted images (T_2WI)................................. 192, 193
Tyrosine hydroxylase (TH) positive neurons 35, 36

U

UK stem cell bank 40
Umbilical cord blood (UCB) cells
    brain homogenates, flow cytometry
        materials 122
        methods 131–132

hematopoietic cells (HPC) 119
immunocytochemistry
    materials 120–121
    methods 124–125
isolation and cryopreservation
    materials 120
    methods 122–124
I.V. transplantation
    materials 121
    methods 129–130
local cell delivery, stereotaxtis
    materials 121
    methods 130–131
migration, injured brain extracts
    materials 121
    methods 127–129
multicolor flow cytometry
    materials 121
    methods 125–126
organotypic hippocampal culture
    materials 122
    methods 131

X

Xenotransplantation 11, 12

Z

Zebrafish. *See Dania rerio*